Understanding Genetics

Russell Torres

STATES
ACADEMIC PRESS
www.statesacademicpress.com

Understanding Genetics
Russell Torres
ISBN: 978-1-63989-538-0 (Hardback)

STATES
ACADEMIC PRESS

Published by States Academic Press,
109 South 5th Street,
Brooklyn, NY 11249, USA

Cataloging-in-Publication Data

Understanding genetics / Russell Torres.
p. cm.
Includes bibliographical references and index.
ISBN 978-1-63989-538-0
1. Genetics. 2. Biology. 3. Biotechnology. I. Torres, Russell.
QH430 .U53 2022
576.5--dc23

For more information regarding States Academic Press and its products, please visit the publisher's website www.statesacademicpress.com

Table of Contents

Preface VII

Chapter 1 **Introduction to Genetics** 1
- a. Components of Genetics 3
- b. Methods in Genetics 52

Chapter 2 **Principles of Genetics** 57
- a. Law of Inheritance 57
- b. Deletion 67
- c. Gene Duplication 73
- d. Chromosome Abnormality 75
- e. Pleiotropy 76
- f. Ploidy 79
- g. Natural Selection 97
- h. Genetic Variation 97
- i. Genetic Drift 102

Chapter 3 **Gene Expression and Regulation** 105
- a. Gene Expression 105
- b. Gene Regulation 110

Chapter 4 **Understanding Genetic Engineering** 120
- a. Cloning 121
- b. Transgenic Technology 124
- c. Gene Therapy 139
- d. Gene Targeting 141
- e. Genetically Modified Organism 152

Chapter 5 **Human Genetics and Genomics** 158
- a. Human Genetics 158
- b. Human Genome 175
- c. Genetic Disorder 188
- d. Gene Therapy 193

Chapter 6 **Various Types of Genetic Disorders** 206
- a. Down Syndrome 207
- b. Cystic Fibrosis 211
- c. Huntington's Disease 216
- d. Duchenne Muscular Dystrophy 218
- e. Sickle Cell Anemia 221
- f. Haemophilia 224

g. Thalassemia 225
h. Fragile X Syndrome 226

Permissions

Index

Preface

The branch of biology which focuses on the study of genes, heredity and genetic variations in various organisms is known as genetics. Some of the various sub fields of genetics involve epigenetics, molecular genetics and population genetics. The functions, variations and distribution of gene structure are studied under this subject within the context of cells, organisms and population. Development and behavior of an organism is often influenced by the genetic processes which work in combination with the environment of an organism. This is also referred to as nature versus nurture in genetics. The gene transcription of a living cell may be turned on or off depending upon its intracellular and extracellular environment. The various sub-fields of genetics along with technological progress that have future implications are glanced at in this book. Some of the diverse topics covered herein address the varied branches that fall under this category. As this field is emerging at a rapid pace, the contents of this book will help the readers understand the modern concepts and applications of the subject.

To facilitate a deeper understanding of the contents of this book a short introduction of every chapter is written below:

Chapter 1- The branch of biology which deals with the study of genes, genetic variations and heredity in organisms is known as genetics. There are various components of genetics such as gene, nucleic acid, nucleotide, chromosome, heredity, genetic code, genotype, etc. This is an introductory chapter which will briefly introduce all these significant components of genetics.

Chapter 2- Principles of genetics include law of inheritance, natural selection, chromosome abnormality, genetic variation, etc. Natural selection is a key mechanism of evolution that describes the differences in the survival and reproduction of individuals having different phenotypes. All these diverse principles of genetics have been carefully analyzed in this chapter.

Chapter 3- The process of formation of a functional gene product by using the information from a gene is termed as gene expression. Gene expression refers to the mechanisms that are used by the cells to regulate the production of specific gene products. This chapter has been carefully written to provide an easy understanding of the varied facets of gene expression and regulation.

Chapter 4- Genetic engineering is a set of technologies that are used to manipulate the genes of an organism. A few of the techniques used in genetic engineering are cloning, transgenic technology, gene therapy, gene targeting, etc. The aim of this chapter is to explore these techniques of genetic engineering which are crucial for a complete understanding of the subject.

Chapter 5- Human genetics is a subfield of genetics that is concerned with the study of the human genome and inheritance in human beings. The complete set of nucleic acid sequence encoded in the DNA of humans is referred to as the human genome. This chapter closely examines the key concepts of human genetics to provide an extensive understanding of the subject.

Chapter 6- The health problems caused by abnormalities in the genome of an individual are known as genetic disorders. These include Down syndrome, cystic fibrosis, Huntington's disease, sickle cell anemia, haemophilia, etc. The topics elaborated in this chapter will help in gaining a better perspective about these genetic disorders.

I would like to share the credit of this book with my editorial team who worked tirelessly on this book. I owe the completion of this book to the never-ending support of my family, who supported me throughout the project.

Russell Torres

1

Introduction to Genetics

The branch of biology which deals with the study of genes, genetic variations and heredity in organisms is known as genetics. There are various components of genetics such as gene, nucleic acid, nucleotide, chromosome, heredity, genetic code, genotype, etc. This is an introductory chapter which will briefly introduce all these significant components of genetics.

Genetics is the study of genes and inheritance in living organisms. This branch of science has a fascinating history, stretching from the 19th century when scientists began to study how organisms inherited traits from their parents, to the present day when we can read the "source code" of living things letter-by-letter. Genetics started out with curiosity about why things are the way things are – why do children resemble one parent more than another? Why do some species resemble each other more closely than others? It has evolved into an almost universal answer handbook for biology. By reading the "source code" or "blueprint" for an organism, scientists today are often able to pinpoint exactly where an organism came from, how it has changed over time, what diseases it might develop, and how its life processes are similar to or different from those of other organisms.

In the 19th century, it was known that offspring resemble their parents – but almost nothing was known about why this happened. Why did some children "take after" one parent, but not the other. Why could plants and animals have offspring that had traits seen in neither parent? Why did some species resemble each other more closely than others? In the 19th century, Gregor Mendel began examining inheritance in a systematic way by breeding pea plants. He tracked several traits of pea plants across several generations, recording what kinds of parents had what kinds off offspring. He successfully derived the mathematics behind dominant and recessive genes – the first empirical evidence that traits really were passed down in some measurable way from parent to offspring.

The image below shows a Punnett square of Mendel's pea plants. The Punnett square was developed by English geneticist Reginald Punnett to visually represent how dominant and recessive traits were passed to offspring. The math yielded by the Punnett square matched the results Mendel found in his hands-on studies of pea plants.

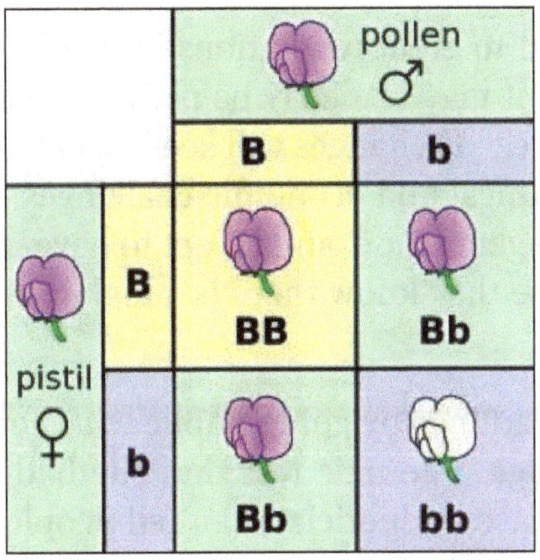

Around the same time, Charles Darwin was writing "The Origin of Species," after examining changes in the traits if island finches during times of drought and plenty. Darwin concluded that finches which had the traits best-suited for survival were most likely to survive to pass those traits on, yielding changes in the traits of the overall population over time. His work, when taken together with Mendel's, began to suggest that all species on Earth might be related to each other, and might have gradually drifted apart by inheriting different traits through natural selection.

From there, the field of genetics advanced slowly. By the early 20th century, scientists using light microscopes powerful enough to see into a cell's nucleus suspected that chromosomes were the seat of genetic information. They were able to connect chromosomal inheritance to trait inheritance, proving that the "instructions" for inherited traits were carried on chromosomes within the nucleus of eukaryotic cells. The next great break in genetics started in the late 20th century, when the technology to read the nucleotide "source code" of the genome began to become available. Since then, the technology has gotten faster, more affordable, and more accurate – allowing scientists to sequence the whole genomes of many organisms and compare them.

Impact of Genetics

The ability to read the "source code of life" has led to a revolution in the way we think about and classify organisms. Prior to the advent of gene sequencing, scientists guessed at organisms' relationships to each other by studying their physical characteristics. Organisms with similar characteristics were often assumed to be related – even though many examples were known of convergent evolution, where two unrelated organisms evolve the same traits separately. With the advent of gene sequencing and molecular genetics- referring to the ability to read the DNA molecule at the molecular level- it became possible to trace descent lineages directly. Scientists can now read a cell's source code and determine at where and roughly when, an organism's genome changed.

As a result, a great deal of material that was taught in schools as recently as ten years ago is now known to be incomplete. Archaea and bacteria – once classified in the same kingdom – are now known to be genetically quite different from each other. Fungi are now known to be more closely related to animals than plants. Many other fantastically weird and fascinating discoveries have come out of the genome revolution – each one bringing us a step closer to understanding what makes us who we are, and how we are interconnected. Gene sequencing has also led to a revolution in the way we think about, diagnose, and treat disease. In many cases, it's now possible to know how likely a person is to get a given disease based on looking at their genome.

Scientists hope that this will lead to great revolutions in medicine in the centuries to come – as medicine catches up to genetics, it may someday be possible to determine what medications will work best on a disease, or what lifestyle changes will keep a person healthy, simply by reading their DNA. This has also led to new ethical and economic challenges. Some women whose genes have certain mutation of the BRCA1/2 gene, for example, opt to have their breasts and ovaries removed even if they are healthy – because they know there is a high chance that they will develop cancer in these organs.

In 2013, Angelina Jolie made headlines by going public with her choice to have her own breasts removed after finding out through a genetic test that she had an 87% chance of someday acquiring breast cancer. In other cases, geneticists can tell people that they will develop a serious disease but do not yet have the tools to stop it from happening. People in families with Huntington's disease, for example, can find out if they have the gene for this devastating and inevitably fatal dementia.

Components of Genetics

Gene

Gene is a unit of hereditary information that occupies a fixed position (locus) on a chromosome. Genes achieve their effects by directing the synthesis of proteins.

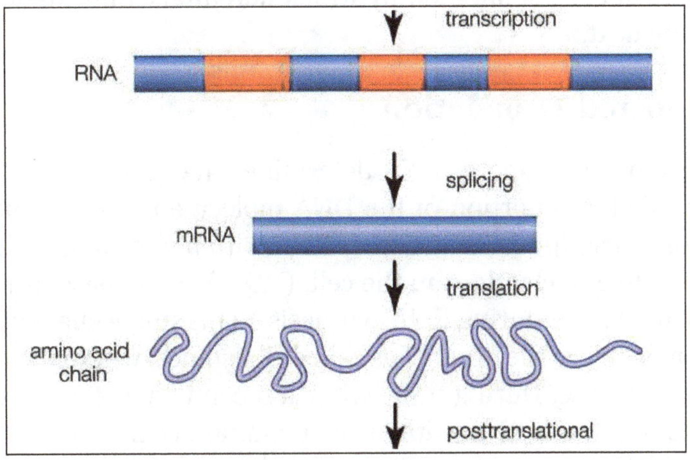

Gene: Intron and exon.

Genes are made up of promoter regions and alternating regions of introns (noncoding sequences) and exons (coding sequences). The production of a functional protein involves the transcription of the gene from DNA into RNA, the removal of introns and splicing together of exons, the translation of the spliced RNA sequences into a chain of amino acids, and the posttranslational modification of the protein molecule.

In eukaryotes (such as animals, plants, and fungi), genes are contained within the cell nucleus. The mitochondria (in animals) and the chloroplasts (in plants) also contain small subsets of genes distinct from the genes found in the nucleus. In prokaryotes (organisms lacking a distinct nucleus, such as bacteria), genes are contained in a single chromosome that is free-floating in the cell cytoplasm. Many bacteria also contain plasmids—extrachromosomal genetic elements with a small number of genes.

The number of genes in an organism's genome (the entire set of chromosomes) varies significantly between species. For example, whereas the human genome contains an estimated 20,000 to 25,000 genes, the genome of the bacterium *Escherichia coli* O157:H7 houses precisely 5,416 genes. Arabidopsis thaliana—the first plant for which a complete genomic sequence was recovered—has roughly 25,500 genes; its genome is one of the smallest known to plants. Among extant independently replicating organisms, the bacterium Mycoplasma genitalium has the fewest number of genes, just 517.

Chemical Structure of Genes

Genes are composed of deoxyribonucleic acid (DNA), except in some viruses, which have genes consisting of a closely related compound called ribonucleic acid (RNA). A DNA molecule is composed of two chains of nucleotides that wind about each other to resemble a twisted ladder. The sides of the ladder are made up of sugars and phosphates, and the rungs are formed by bonded pairs of nitrogenous bases. These bases are adenine (A), guanine (G), cytosine (C), and thymine (T). An A on one chain bonds to a T on the other (thus forming an A–T ladder rung); similarly, a C on one chain bonds to a G on the other. If the bonds between the bases are broken, the two chains unwind, and free nucleotides within the cell attach themselves to the exposed bases of the now-separated chains. The free nucleotides line up along each chain according to the base-pairing rule—A bonds to T, C bonds to G. This process results in the creation of two identical DNA molecules from one original and is the method by which hereditary information is passed from one generation of cells to the next.

Gene Transcription and Translation

The sequence of bases along a strand of DNA determines the genetic code. When the product of a particular gene is needed, the portion of the DNA molecule that contains that gene will split. Through the process of transcription, a strand of RNA with bases complementary to those of the gene is created from the free nucleotides in the cell. (RNA has the base uracil [U] instead of thymine, so A and U form base pairs during RNA synthesis.) This single chain of RNA, called messenger RNA (mRNA), then passes to the organelles called ribosomes, where the process of translation, or protein synthesis, takes place. During translation, a second type of RNA, transfer RNA (tRNA), matches up the nucleotides on mRNA with specific amino acids. Each set of three nucleotides codes for one amino acid. The series of amino acids built according to the sequence of nucleotides forms a polypeptide chain; all proteins are made from one or more linked polypeptide chains.

Experiments conducted in the 1940s indicated one gene being responsible for the assembly of one enzyme, or one polypeptide chain. This is known as the one gene–one enzyme hypothesis. However, since this discovery, it has been realized that not all genes encode an enzyme and that some enzymes are made up of several short polypeptides encoded by two or more genes.

Gene Regulation

Experiments have shown that many of the genes within the cells of organisms are inactive much or even all of the time. Thus, at any time, in both eukaryotes and prokaryotes, it seems that a gene can be switched on or off. The regulation of genes between eukaryotes and prokaryotes differs in important ways.

The process by which genes are activated and deactivated in bacteria is well characterized. Bacteria have three types of genes: structural, operator, and regulator. Structural genes code for the synthesis of specific polypeptides. Operator genes contain the code necessary to begin the process of transcribing the DNA message of one or more structural genes into mRNA. Thus, structural genes are linked to an operator gene in a functional unit called an operon. Ultimately, the activity of the operon is controlled by a regulator gene, which produces a small protein molecule called a repressor. The repressor binds to the operator gene and prevents it from initiating the synthesis of the protein called for by the operon. The presence or absence of certain repressor molecules determines whether the operon is off or on. As mentioned, this model applies to bacteria.

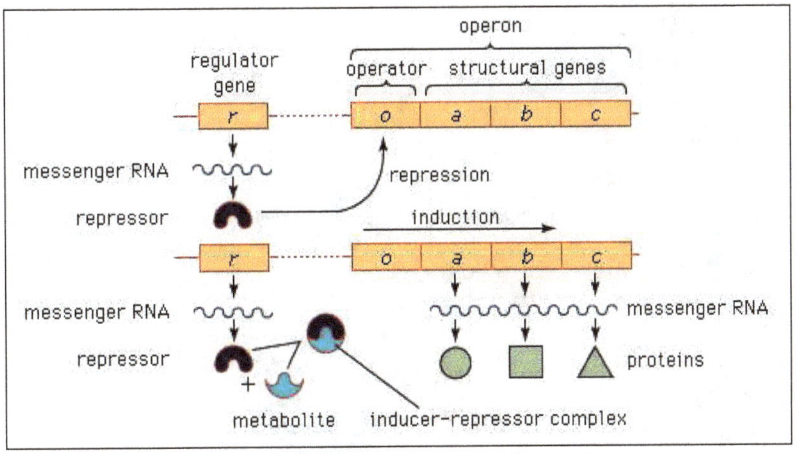

Model of the operon and its relation to the regulator gene.

The genes of eukaryotes, which do not have operons, are regulated independently. The series of events associated with gene expression in higher organisms involves multiple levels of regulation and is often influenced by the presence or absence of molecules called transcription factors. These factors influence the fundamental level of gene control, which is the rate of transcription, and may function as activators or enhancers. Specific transcription factors regulate the production of RNA from genes at certain times and in certain types of cells. Transcription factors often bind to the promoter, or regulatory region, found in the genes of higher organisms. Following transcription, introns (noncoding nucleotide sequences) are excised from the primary transcript through processes known as editing and splicing. The result of these processes is a functional strand of mRNA. For most genes this is a routine step in the production of mRNA, but in some genes there are multiple ways to splice the primary transcript, resulting in different mRNAs, which in turn result in different proteins. Some genes also are controlled at the translational and posttranslational levels.

Gene Mutations

Mutations occur when the number or order of bases in a gene is disrupted. Nucleotides can be deleted, doubled, rearranged, or replaced, each alteration having a particular effect. Mutation generally has little or no effect, but, when it does alter an organism, the change may be lethal or cause disease. A beneficial mutation will rise in frequency within a population until it becomes the norm.

Nucleic Acid

Nucleic acid is the naturally occurring chemical compound that is capable of being broken down to yield phosphoric acid, sugars, and a mixture of organic bases (purines and pyrimidines). Nucleic acids are the main information-carrying molecules of the cell, and, by directing the process of protein synthesis, they determine the inherited characteristics of every living thing. The two main classes of nucleic acids are deoxyribonucleic acid (DNA) and ribonucleic acid (RNA). DNA is the master blueprint for life and constitutes the genetic material in all free-living organisms and most viruses. RNA is the genetic material of certain viruses, but it is also found in all living cells, where it plays an important role in certain processes such as the making of proteins.

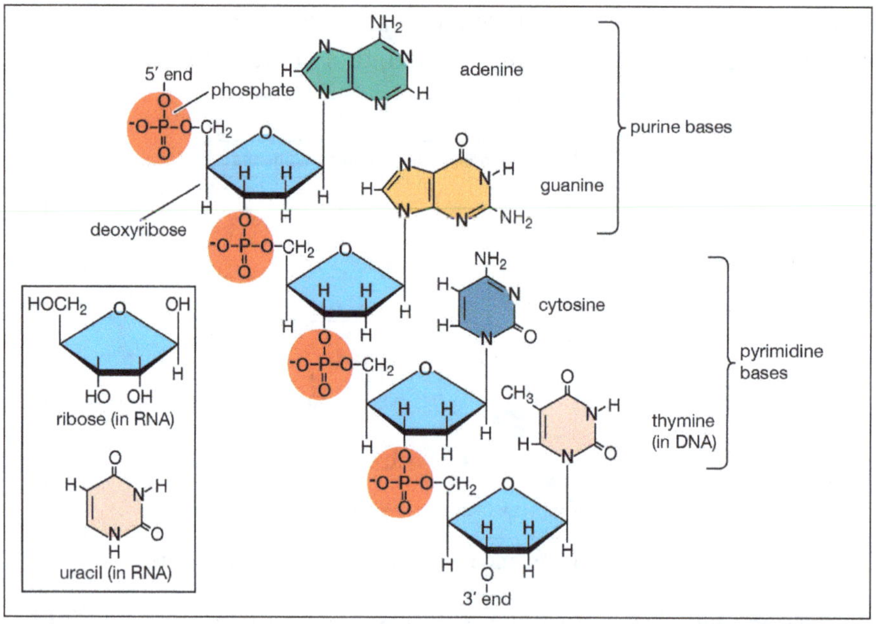

Polynucleotide chain of deoxyribonucleic acid (DNA): Portion of polynucleotide chain of deoxyribonucleic acid (DNA). The inset shows the corresponding pentose sugar and pyrimidine base in ribonucleic acid (RNA).

Nucleotides: Building Blocks of Nucleic Acids

Basic Structure

Nucleic acids are polynucleotides—that is, long chainlike molecules composed of a series of nearly identical building blocks called nucleotides. Each nucleotide consists of a nitrogen-containing aromatic base attached to a pentose (five-carbon) sugar, which is in turn attached to a phosphate group. Each nucleic acid contains four of five possible nitrogen-containing bases: adenine (A), guanine (G), cytosine (C), thymine (T), and uracil (U). A and G are categorized as purines, and C, T, and U are collectively called pyrimidines. All nucleic acids contain the bases A, C, and G; T, however, is found only in DNA, while U is found in RNA. The pentose sugar in DNA (2′-deoxyribose) differs from the sugar in RNA (ribose) by the absence of a hydroxyl group (−OH) on the 2′ carbon of the sugar ring. Without an attached phosphate group, the sugar attached to one of the bases is known as a nucleoside. The phosphate group connects successive sugar residues by bridging the 5′-hydroxyl group on one sugar to the 3′-hydroxyl group of the next sugar in the chain. These nucleoside linkages are called phosphodiester bonds and are the same in RNA and DNA.

Biosynthesis and Degradation

Nucleotides are synthesized from readily available precursors in the cell. The ribose phosphate portion of both purine and pyrimidine nucleotides is synthesized from glucose via the pentose phosphate pathway. The six-atom pyrimidine ring is synthesized first and subsequently attached to the ribose phosphate. The two rings in purines are synthesized while attached to the ribose phosphate during the assembly of adenine or guanine nucleosides. In both cases the end product is a nucleotide carrying a phosphate attached to the 5′ carbon on the sugar. Finally, a specialized enzyme called a kinase adds two phosphate groups using adenosine triphosphate (ATP) as the phosphate donor to form ribonucleoside triphosphate, the immediate precursor of RNA. For DNA, the

2′-hydroxyl group is removed from the ribonucleoside diphosphate to give deoxyribonucleoside diphosphate. An additional phosphate group from ATP is then added by another kinase to form a deoxyribonucleoside triphosphate, the immediate precursor of DNA.

During normal cell metabolism, RNA is constantly being made and broken down. The purine and pyrimidine residues are reused by several salvage pathways to make more genetic material. Purine is salvaged in the form of the corresponding nucleotide, whereas pyrimidine is salvaged as the nucleoside.

Deoxyribonucleic Acid (DNA)

DNA is a polymer of the four nucleotides A, C, G, and T, which are joined through a backbone of alternating phosphate and deoxyribose sugar residues. These nitrogen-containing bases occur in complementary pairs as determined by their ability to form hydrogen bonds between them. A always pairs with T through two hydrogen bonds, and G always pairs with C through three hydrogen bonds. The spans of A:T and G:C hydrogen-bonded pairs are nearly identical, allowing them to bridge the sugar-phosphate chains uniformly. This structure, along with the molecule's chemical stability, makes DNA the ideal genetic material. The bonding between complementary bases also provides a mechanism for the replication of DNA and the transmission of genetic information.

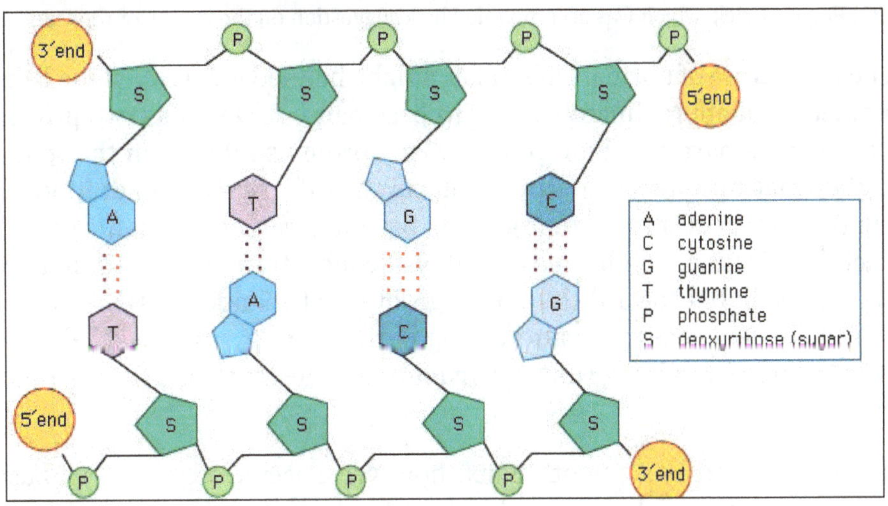

DNA structure.

DNA structure, showing the nucleotide bases cytosine (C), thymine (T), adenine (A), and guanine (G) linked to a backbone of alternating phosphate (P) and deoxyribose sugar (S) groups. Two sugar-phosphate chains are paired through hydrogen bonds between A and T and between G and C, thus forming the twin-stranded double helix of the DNA molecule.

Chemical Structure

In 1953 James D. Watson and Francis H.C. Crick proposed a three-dimensional structure for DNA based on low-resolution X-ray crystallographic data and on Erwin Chargaff's observation that, in naturally occurring DNA, the amount of T equals the amount of A and the amount of G equals the amount of C. Watson and Crick, who shared a Nobel Prize in 1962 for their efforts, postulated that two strands of polynucleotides coil around each other, forming a double helix. The two strands, though identical, run in opposite directions as determined by the orientation of the 5′ to 3′ phosphodiester bond. The sugar-phosphate chains run along the outside of the helix, and the bases lie on the inside, where they are linked to complementary bases on the other strand through hydrogen bonds.

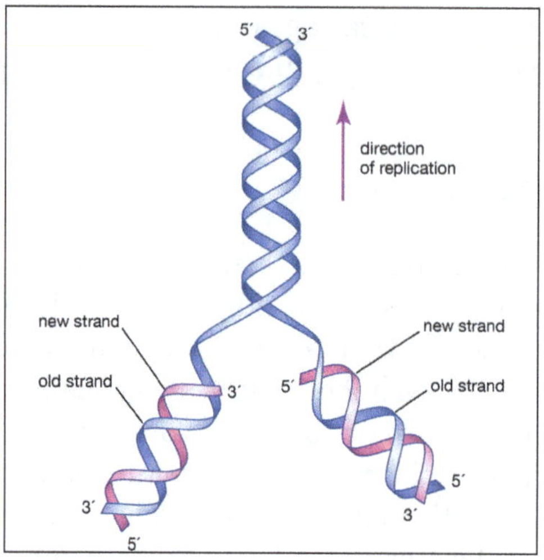

Initial proposal of DNA structure: The initial proposal of the structure of DNA by James Watson and Francis Crick, which was accompanied by a suggestion on the means of replication.

The double helical structure of normal DNA takes a right-handed form called the B-helix. The helix makes one complete turn approximately every 10 base pairs. B-DNA has two principal grooves, a wide major groove and a narrow minor groove. Many proteins interact in the space of the major groove, where they make sequence-specific contacts with the bases. In addition, a few proteins are known to make contacts via the minor groove. Several structural variants of DNA are known. In A-DNA, which forms under conditions of high salt concentration and minimal water, the base pairs are tilted and displaced toward the minor groove. Left-handed Z-DNA forms most readily in strands that contain sequences with alternating purines and pyrimidines. DNA can form triple helices when two strands containing runs of pyrimidines interact with a third strand containing a run of purines.

B-DNA is generally depicted as a smooth helix; however, specific sequences of bases can distort the otherwise regular structure. For example, short tracts of A residues interspersed with short sections of general sequence result in a bent DNA molecule. Inverted base sequences, on the other hand, produce cruciform structures with four-way junctions that are similar to recombination intermediates. Most of these alternative DNA structures have only been characterized in the laboratory, and their cellular significance is unknown.

Biological Structures

Naturally occurring DNA molecules can be circular or linear. The genomes of single-celled bacteria and archaea (the prokaryotes), as well as the genomes of mitochondria and chloroplasts (certain functional structures within the cell), are circular molecules. In addition, some bacteria and archaea have smaller circular DNA molecules called plasmids that typically contain only a few genes. Many plasmids are readily transmitted from one cell to another. For a typical bacterium, the genome that encodes all of the genes of the organism is a single contiguous circular molecule that contains a half million to five million base pairs. The genomes of most eukaryotes and some prokaryotes contain linear DNA molecules called chromosomes. Human DNA, for example, consists of 23 pairs of linear chromosomes containing three billion base pairs.

In all cells, DNA does not exist free in solution but rather as a protein-coated complex called chromatin. In prokaryotes, the loose coat of proteins on the DNA helps to shield the negative charge of the phosphodiester backbone. Chromatin also contains proteins that control gene expression and determine the characteristic shapes of chromosomes. In eukaryotes, a section of DNA between 140 and 200 base pairs long winds around a discrete set of eight positively charged proteins called a histone, forming a spherical structure called the nucleosome. Additional histones are wrapped by successive sections of DNA, forming a series of nucleosomes like beads on a string. Transcription and replication of DNA is more complicated in eukaryotes because the nucleosome complexes have to be at least partially disassembled for the processes to proceed effectively.

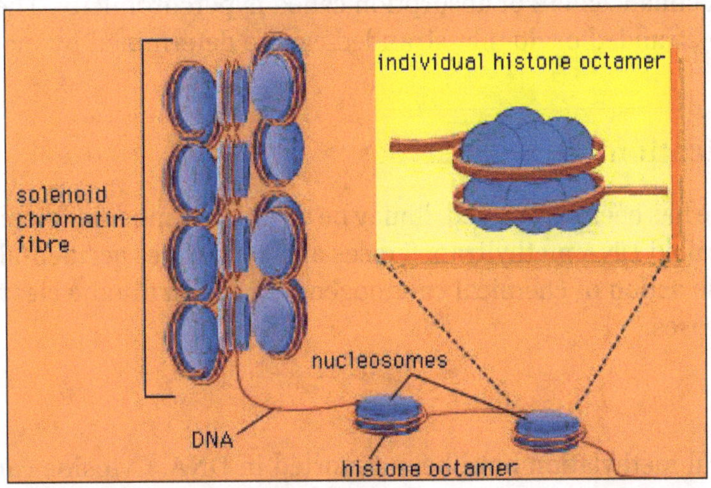

DNA wrapped around clusters of histone proteins to form nucleosomes, which are coiled to form solenoids, the basis of the chromatin fibre that makes up chromosomes.

Most prokaryote viruses contain linear genomes that typically are much shorter and contain only the genes necessary for viral propagation. Bacterial viruses called bacteriophages (or phages) may contain both linear and circular forms of DNA. For instance, the genome of bacteriophage λ (lambda), which infects the bacterium *Escherichia coli*, contains 48,502 base pairs and can exist as a linear molecule packaged in a protein coat. The DNA of phage λ can also exist in a circular form that is able to integrate into the circular genome of the host bacterial cell. Both circular and linear genomes are found among eukaryotic viruses, but they more commonly use RNA as the genetic material.

Biochemical Properties

Denaturation

The strands of the DNA double helix are held together by hydrogen bonding interactions between the complementary base pairs. Heating DNA in solution easily breaks these hydrogen bonds, allowing the two strands to separate—a process called denaturation or melting. The two strands may reassociate when the solution cools, reforming the starting DNA duplex—a process called renaturation or hybridization. These processes form the basis of many important techniques for manipulating DNA. For example, a short piece of DNA called an oligonucleotide can be used to test whether a very long DNA sequence has the complementary sequence of the oligonucleotide embedded within it. Using hybridization, a single-stranded DNA molecule can capture complementary sequences from any source. Single strands from RNA can also reassociate. DNA and RNA single strands can form hybrid molecules that are even more stable than double-stranded DNA. These molecules form the basis of a technique that is used to purify and characterize messenger RNA (mRNA) molecules corresponding to single genes.

Ultraviolet Absorption

DNA melting and reassociation can be monitored by measuring the absorption of ultraviolet (UV) light at a wavelength of 260 nanometres (billionths of a metre). When DNA is in a double-stranded conformation, absorption is fairly weak, but when DNA is single-stranded, the unstacking of the bases leads to an enhancement of absorption called hyperchromicity. Therefore, the extent to which DNA is single-stranded or double-stranded can be determined by monitoring UV absorption.

Chemical Modification

After a DNA molecule has been assembled, it may be chemically modified—sometimes deliberately by special enzymes called DNA methyltransferases and sometimes accidentally by oxidation, ionizing radiation, or the action of chemical carcinogens. DNA can also be cleaved and degraded by enzymes called nucleases.

Methylation

Three types of natural methylation have been reported in DNA. Cytosine can be modified either on the ring to form 5-methylcytosine or on the exocyclic amino group to form N^4-methylcytosine. Adenine may be modified to form N^6-methyladenine. N^4-methylcytosine and N^6-methyladenine are found only in bacteria and archaea, whereas 5-methylcytosine is widely distributed. Special enzymes called DNA methyltransferases are responsible for this methylation; they recognize specific sequences within the DNA molecule so that only a subset of the bases is modified. Other methylations of the bases or of the deoxyribose are sometimes induced by carcinogens. These usually lead to mispairing of the bases during replication and have to be removed if they are not to become mutagenic.

Natural methylation has many cellular functions. In bacteria and archaea, methylation forms an essential part of the immune system by protecting DNA molecules from fragmentation by restriction endonucleases. In some organisms, methylation helps to eliminate incorrect base sequences introduced during DNA replication. By marking the parental strand with a methyl group, a cellular mechanism known as the mismatch repair system distinguishes between the newly replicated strand where the errors occur and the correct sequence on the template strand. In higher eukaryotes, 5-methylcytosine controls many cellular phenomena by preventing DNA transcription.

Methylation is also believed to signal imprinting, a process whereby some genes inherited from one parent are selectively inactivated. Correct methylation may also repress or activate key genes that control embryonic development. On the other hand, 5-methylcytosine is potentially mutagenic because thymine produced during the methylation process converts C:G pairs to T:A pairs. In mammals, methylation takes place selectively within the dinucleotide sequence CG—a rare sequence, presumably because it has been lost by mutation. In many cancers, mutations are found in key genes at CG dinucleotides.

Nucleases

Nucleases are enzymes that hydrolytically cleave the phosphodiester backbone of DNA. Endonucleases cleave in the middle of chains, while exonucleases operate selectively by degrading from the end of the chain. Nucleases that act on both single- and double-stranded DNA are known. Restriction endonucleases are a special class that recognize and cleave specific sequences in DNA. Type II restriction endonucleases always cleave at or near their recognition sites. They produce small, well-defined fragments of DNA that help to characterize genes and genomes and that produce recombinant DNAs. Fragments of DNA produced by restriction endonucleases can be moved from one organism to another. In this way it has been possible to express proteins such as human insulin in bacteria.

Mutation

Chemical modification of DNA can lead to mutations in the genetic material. Anions such as bisulfite can deaminate cytosine to form uracil, changing the genetic message by causing C-to-T transitions. Exposure to acid causes the loss of purine residues, though specific enzymes exist in cells to repair these lesions. Exposure to UV light can cause adjacent pyrimidines to dimerize, while oxidative damage from free radicals or strong oxidizing agents can cause a variety of lesions that are mutagenic if not repaired. Halogens such as chlorine and bromine react directly with uracil, adenine, and guanine, giving substituted bases that are often mutagenic. Similarly, nitrous acid reacts with primary amine groups—for example, converting adenosine into inosine—which then leads to changes in base pairing and mutation. Many chemical mutagens, such as chlorinated hydrocarbons and nitrites, owe their toxicity to the production of halides and nitrous acid during their metabolism in the body.

Supercoiling

Circular DNA molecules such as those found in plasmids or bacterial chromosomes can adopt many different topologies. One is active supercoiling, which involves the cleavage of one DNA strand, its winding one or more turns around the complementary strand, and then the resealing of the molecule. Each complete rotation leads to the introduction of one supercoiled turn in the DNA, a process that can continue until the DNA is fully wound and collapses on itself in a tight ball. Reversal is also possible. Special enzymes called gyrases and topoisomerases catalyze the winding and relaxation of supercoiled DNA. In the linear chromosomes of eukaryotes, the DNA is usually tightly constrained at various points by proteins, allowing the intervening stretches to be supercoiled. This property is partially responsible for the great compaction of DNA that is necessary to fit it within the confines of the cell. The DNA in one human cell would have an extended length of between two and three metres, but it is packed very tightly so that it can fit within a human cell nucleus that is 10 micrometres in diameter.

Sequence Determination

Methods to determine the sequences of bases in DNA were pioneered in the 1970s by Frederick Sanger and Walter Gilbert, whose efforts won them a Nobel Prize in 1980. The Gilbert-Maxam method relies on the different chemical reactivities of the bases, while the Sanger method is based on enzymatic synthesis of DNA in vitro. Both methods measure the distance from a fixed point on DNA to each occurrence of a particular base—A, C, G, or T. DNA fragments obtained from a series of reactions are separated according to length in four "lanes" by gel electrophoresis. Each lane corresponds to a unique base, and the sequence is read directly from the gel. The Sanger method has now been automated using fluorescent dyes to label the DNA, and a single machine can produce tens of thousands of DNA base sequences in a single run.

Ribonucleic Acid (RNA)

RNA is a single-stranded nucleic acid polymer of the four nucleotides A, C, G, and U joined through a backbone of alternating phosphate and ribose sugar residues. It is the first intermediate in converting the information from DNA into proteins essential for the working of a cell. Some RNAs also serve direct roles in cellular metabolism. RNA is made by copying the base sequence of a section of double-stranded DNA, called a gene, into a piece of single-stranded nucleic acid. This process, called transcription is catalyzed by an enzyme called RNA polymerase.

Chemical Structure

Whereas DNA provides the genetic information for the cell and is inherently quite stable, RNA has many roles and is much more reactive chemically. RNA is sensitive to oxidizing agents such as periodate that lead to opening of the 3′-terminal ribose ring. The 2′-hydroxyl group on the ribose ring is a major cause of instability in RNA, because the presence of alkali leads to rapid cleavage of the phosphodiester bond linking ribose and phosphate groups. In general, this instability is not a significant problem for the cell, because RNA is constantly being synthesized and degraded.

Interactions between the nitrogen-containing bases differ in DNA and RNA. In DNA, which is usually double-stranded, the bases in one strand pair with complementary bases in a second DNA strand. In RNA, which is usually single-stranded, the bases pair with other bases within the same molecule, leading to complex three-dimensional structures. Occasionally, intermolecular RNA/RNA duplexes do form, but they form a right-handed A-type helix rather than the B-type DNA helix. Depending on the amount of salt present, either 11 or 12 base pairs are found in each turn of the helix. Helices between RNA and DNA molecules also form; these adopt the A-type conformation and are more stable than either RNA/RNA or DNA/DNA duplexes. Such hybrid duplexes are important species in biology, being formed when RNA polymerase transcribes DNA into mRNA for protein synthesis and when reverse transcriptase copies a viral RNA genome such as that of the human immunodeficiency virus (HIV).

Single-stranded RNAs are flexible molecules that form a variety of structures through internal base pairing and additional non-base pair interactions. They can form hairpin loops such as those found in transfer RNA (tRNA), as well as longer-range interactions involving both the bases and the phosphate residues of two or more nucleotides. This leads to compact three-dimensional structures. Most of these structures have been inferred from biochemical data, since few crystallographic images are available for RNA molecules. In some types of RNA, a large number of bases are modified after the RNA is transcribed. More than 90 different modifications have been documented, including extensive methylations and a wide variety of substitutions around the ring. In some cases these modifications are known to affect structure and are essential for function.

Types of RNA

Messenger RNA (mRNA)

Messenger RNA (mRNA) delivers the information encoded in one or more genes from the DNA to the ribosome, a specialized structure, or organelle, where that information is decoded into a protein. In prokaryotes, mRNAs contain an exact transcribed copy of the original DNA sequence with a terminal 5′-triphosphate group and a 3′-hydroxyl residue. In eukaryotes the mRNA molecules are more elaborate. The 5′-triphosphate residue is further esterified, forming a structure called a cap. At the 3′ ends, eukaryotic mRNAs typically contain long runs of adenosine residues (polyA) that are not encoded in the DNA but are added enzymatically after transcription. Eukaryotic mRNA molecules are usually composed of small segments of the original gene and are generated by a process of cleavage and rejoining from an original precursor RNA (pre-mRNA) molecule, which is an exact copy of the gene. In general, prokaryotic mRNAs are degraded very rapidly, whereas the cap structure and the polyA tail of eukaryotic mRNAs greatly enhance their stability.

Ribosomal RNA (rRNA)

Ribosomal RNA (rRNA) molecules are the structural components of the ribosome. The rRNAs form extensive secondary structures and play an active role in recognizing conserved portions of mRNAs and tRNAs. They also assist with the catalysis of protein synthesis. In the prokaryote *E. coli*, seven copies of the rRNA genes synthesize about 15,000 ribosomes per cell. In eukaryotes the numbers are much larger. Anywhere from 50 to 5,000 sets of rRNA genes and as many as 10 million ribosomes may be present in a single cell. In eukaryotes these rRNA genes are looped out of the main chromosomal fibres and coalesce in the presence of proteins to form an organelle called the nucleolus. The nucleolus is where the rRNA genes are transcribed and the early assembly of ribosomes takes place.

Transfer RNA (tRNA)

Transfer RNA (tRNA) carries individual amino acids into the ribosome for assembly into the growing polypeptide chain. The tRNA molecules contain 70 to 80 nucleotides and fold into a characteristic cloverleaf structure. Specialized tRNAs exist for each of the 20 amino acids needed for protein synthesis, and in many cases more than one tRNA for each amino acid is present. The nucleotide sequence is converted into a protein sequence by translating each three-base sequence (called a codon) with a specific protein. The 61 codons used to code amino acids can be read by many fewer than 61 distinct tRNAs. In *E. coli* a total of 40 different tRNAs are used to translate the 61 codons. The amino acids are loaded onto the tRNAs by specialized enzymes called aminoacyl tRNA synthetases, usually with one synthetase for each amino acid. However, in some organisms, less than the full complement of 20 synthetases are required because some amino acids, such as glutamine and asparagine, can be synthesized on their respective tRNAs. All tRNAs adopt similar structures because they all have to interact with the same sites on the ribosome.

Ribozymes

Not all catalysis within the cell is carried out exclusively by proteins. Thomas Cech and Sidney Altman, jointly awarded a Nobel Prize in 1989, discovered that certain RNAs, now known as ribozymes, showed enzymatic activity. Cech showed that a noncoding sequence (intron) in the small subunit rRNA of protozoans, which had to be removed before the rRNA was functional, can excise itself from a much longer precursor RNA molecule and rejoin the two ends in an autocatalytic reaction. Altman showed that the RNA component of an RNA protein complex called ribonuclease P can cleave a precursor tRNA to generate a mature tRNA. In addition to self-splicing RNAs similar to the one discovered by Cech, artificial RNAs have been made that show a variety of catalytic reactions. It is now widely held that there was a stage during evolution when only RNA catalyzed and stored genetic information. This period, sometimes called "the RNA world," is believed to have preceded the function of DNA as genetic material.

Antisense RNAs

Most antisense RNAs are synthetically modified derivatives of RNA or DNA with potential therapeutic value. In nature, antisense RNAs contain sequences that are the complement of the normal coding sequences found in mRNAs (also called sense RNAs). Like mRNAs, antisense RNAs are single-stranded, but they cannot be translated into protein. They can inactivate their complementary mRNA by forming a double-stranded structure that blocks the translation of the base sequence. Artificially introducing antisense RNAs into cells selectively inactivates genes by interfering with normal RNA metabolism.

Viral Genomes

Many viruses use RNA for their genetic material. This is most prevalent among eukaryotic viruses, but a few prokaryotic RNA viruses are also known. Some common examples include poliovirus, human immunodeficiency virus (HIV), and influenza virus, all of which affect humans, and tobacco mosaic virus, which infects plants. In some viruses the entire genetic material is encoded in a single RNA molecule, while in the segmented RNA viruses several RNA molecules may be present. Many RNA viruses such as HIV use a specialized enzyme called reverse transcriptase that permits replication of the virus through a DNA intermediate. In some cases this DNA intermediate becomes integrated into the host chromosome during infection; the virus then exists in a dormant state and effectively evades the host immune system.

Other RNAs

Many other small RNA molecules with specialized functions are present in cells. For example, small nuclear RNAs (snRNAs) are involved in RNA splicing and other small RNAs that form part of the enzymes telomerase or ribonuclease P are part of ribonucleoprotein particles. The RNA component of telomerase contains a short sequence that serves as a template for the addition of small strings of oligonucleotides at the ends of eukaryotic chromosomes. Other RNA molecules serve as guide RNAs for editing, or they are complementary to small sections of rRNA and either direct the positions at which methyl groups need to be added or mark U residues for conversion to the isomer pseudouridine.

RNA Processing

Cleavage

Following synthesis by transcription, most RNA molecules are processed before reaching their final form. Many rRNA molecules are cleaved from much larger transcripts and may also be methylated or enzymatically modified. In addition, tRNAs are usually formed as longer precursor molecules that are cleaved by ribonuclease P to generate the mature 5′ end and often have extra residues added to their 3′ end to form the sequence CCA. The hydroxyl group on the ribose ring of the terminal A of the 3′-CCA sequence acts as the amino acid acceptor necessary for the function of RNA in protein building.

Splicing

In prokaryotes the protein coding sequence occupies one continuous linear segment of DNA. However, in eukaryotic genes the coding sequences are frequently "split" in the genome—a discovery reached independently in the 1970s by Richard J. Roberts and Phillip A. Sharp, whose work won them a Nobel Prize in 1993. The segments of DNA or RNA coding for protein are called exons, and the noncoding regions separating the exons are called introns. Following transcription, these coding sequences must be joined together before the mRNAs can function. The process of removal of the introns and subsequent rejoining of the exons is called RNA splicing. Each intron is removed in a separate series of reactions by a complicated piece of enzymatic machinery called a spliceosome. This machinery consists of a number of small nuclear ribonucleoprotein particles (snRNPs) that contain small nuclear RNAs (snRNAs).

RNA Editing

Some RNA molecules, particularly those in protozoan mitochondria, undergo extensive editing following their initial synthesis. During this editing process, residues are added or deleted by a posttranscriptional mechanism under the influence of guide RNAs. In some cases as much as 40 percent of the final RNA molecule may be derived by this editing process, rather than being coded directly in the genome. Some examples of editing have also been found in mRNA molecules, but these appear much more limited in scope.

Nucleic Acid Metabolism

DNA Metabolism

Replication, repair, and recombination—the three main processes of DNA metabolism—are carried out by specialized machinery within the cell. DNA must be replicated accurately in order to ensure the integrity of the genetic code. Errors that creep in during replication or because of damage after replication must be repaired. Finally, recombination between genomes is an important mechanism to provide variation within a species and to assist the repair of damaged DNA. The details of each process have been worked out in prokaryotes, where the machinery is more streamlined, simpler, and more amenable to study. Many of the basic principles appear to be similar in eukaryotes.

Replication

Basic Mechanisms

DNA replication is a semiconservative process in which the two strands are separated and new complementary strands are generated independently, resulting in two exact copies of the original DNA molecule. Each copy thus contains one strand that is derived from the parent and one newly synthesized strand. Replication begins at a specific point on a chromosome called an origin, proceeds in both directions along the strand, and ends at a precise point. In the case of circular chromosomes, the end is reached automatically when the two extending chains meet, at which point specific proteins join the strands. DNA polymerases cannot initiate replication at the end of a DNA strand; they can only extend preexisting oligonucleotide fragments called primers. Therefore, in linear chromosomes, special mechanisms initiate and terminate DNA synthesis to avoid loss of information. The initiation of DNA synthesis is usually preceded by synthesis of a short RNA primer by a specialized RNA polymerase called primase. Following DNA replication, the initiating primer RNAs are degraded.

The two DNA strands are replicated in different fashions dictated by the direction of the phosphodiester bond. The leading strand is replicated continuously by adding individual nucleotides to the 3′ end of the chain. The lagging strand is synthesized in a discontinuous manner by laying down short RNA primers and then filling the gaps by DNA polymerase, such that the bases are always added in the 5′ to 3′ direction. The short RNA fragments made during the copying of the lagging strand are degraded when no longer needed. The two newly synthesized DNA segments are joined by an enzyme called DNA ligase. In this way, replication can proceed in both directions, with two leading strands and two lagging strands proceeding outward from the origin.

Enzymes of Replication

DNA polymerase adds single nucleotides to the 3′ end of either an RNA or a DNA molecule. In the prokaryote *E. coli*, there are three DNA polymerases; one is responsible for chromosome replication, and the other two are involved in the resynthesis of DNA during damage repair. DNA polymerases of eukaryotes are even more complicated. In human cells, for instance, more than five different DNA polymerases have been characterized. Separate polymerases catalyze the synthesis of the leading and lagging strands in human cells, and a separate polymerase is responsible for replication of mitochondrial DNA. The other polymerases are involved in the repair of DNA damage.

A number of other proteins are also essential for replication. Proteins called DNA helicases help to separate the two strands of DNA, and single-stranded DNA binding proteins stabilize them during opening prior to being copied. The opening of the DNA helix introduces considerable strain in the form of supercoiling, a movement that is subsequently relaxed by enzymes called topoisomerases. A special RNA polymerase called primase synthesizes the primers needed at the origin to begin transcription, and DNA ligase seals the nicks formed between individual fragments.

The ends of linear eukaryotic chromosomes are marked by special sequences called telomeres that are synthesized by a special DNA polymerase called telomerase. This enzyme contains an RNA component that serves as a template for the exact sequence found at the ends of chromosomes. Multiple copies of a short sequence within the telomerase-associated RNA are made and added to the telomere ends. This has the effect of preventing shortening of the DNA chain that would otherwise occur during replication.

Single-stranded viral genomes, mitochondrial genomes, and some viral genomes are replicated in specialized ways. Several viruses such as adenovirus use a nucleotide covalently bound to a protein as a primer, and the protein remains covalently bound to the DNA after replication. Many single-stranded viruses use a rolling circle mechanism of replication whereby a double-stranded copy of the virus is first made. The replicating machinery then copies the nonviral strand in a continuous fashion, generating long single-stranded DNA from which full-length viral DNA strands are excised by specialized nucleases.

Recombination

Recombination is the principal mechanism through which variation is introduced into populations. For example, during meiosis, the process that produces sex cells (sperm or eggs), homologous chromosomes—one derived from the mother and the equivalent from the father—become paired, and recombination, or crossing-over, takes place. The two DNA molecules are fragmented, and similar segments of the chromosome are shuffled to produce two new chromosomes, each being a mosaic of the originals. The pair separates so that each sperm or egg receives just one of the shuffled chromosomes. When sperm and egg fuse, the normal set of two copies of each chromosome is restored.

There are two forms of recombination, general and site-specific. General recombination typically involves cleavage and rejoining at identical or very similar sequences. In site-specific recombination, cleavage takes place at a specific site into which DNA is usually inserted. General recombination occurs among viruses during infection, in bacteria during conjugation, during transformation whereby DNA is directly introduced into cells, and during some types of repair processes. Site-specific recombination is frequently involved in the parasitic distribution of DNA segments throughout genomes. Many viruses, as well as special segments of DNA called transposons, rely on site-specific recombination to multiply and spread.

General Recombination

General recombination, also called homologous recombination, involves two DNA molecules that have long stretches of similar base sequences. The DNA molecules are nicked to produce single strands; these subsequently invade the other duplex, where base pairing leads to a four-stranded DNA structure. The cruciform junction within this structure is called a Holliday junction, named after Robin Holliday, who proposed the original model for homologous recombination in 1964. The Holliday junction travels along the DNA duplex by "unzipping" one strand and reforming the hydrogen bonds on the second strand. Following this branch migration, the two duplexes can be nicked again, allowing them to separate. Finally, the nicks are repaired by DNA ligase. The result is two DNA duplexes in which the segment between the two nicks has been replaced. The enzymes involved in recombination have been characterized best in the prokaryote E. coli. A key enzyme is

RecA, which catalyzes the strand invasion process. RecA coats single-stranded DNA and facilitates its pairing with a double-stranded DNA molecule containing the same sequence, which produces a loop structure.

Another protein, known as RecBC, is important for the recombination process. Functioning at free ends of DNA, RecBC catalyzes an unwinding-rewinding reaction as it traverses the length of the molecule. Since unwinding is faster than rewinding, a loop is produced behind the enzyme that facilitates subsequent pairing with another DNA molecule. A number of other proteins are also important for recombination, including single-stranded DNA binding proteins that stabilize single-stranded DNA, DNA polymerase to repair any gaps that might be formed, and DNA ligase to reseal the nicks after recombination is complete. The details of eukaryotic recombination are expected to parallel those found in E. coli, although the highly compact chromatin structure in eukaryotes makes the process more complicated.

It is important to note that the initial product of recombination between two regions of DNA that are similar but not identical will be a "heteroduplex"—that is, a molecule in which mismatched bases will be present at some positions in the helix. Thus, in the specialized recombination that takes place during meiosis, one round of replication is necessary before the mosaic chromosomes produced by recombination are properly matched. Enzymes are present in cells that specifically recognize and repair mismatches, so that the initial products of recombination can sometimes be repaired before they are replicated. In such cases the final products of replication will not be true reciprocal events, but rather one of the original parental molecules will appear to have been maintained to the exclusion of the other—a process called gene conversion.

Recombination also functions occasionally to repair lesions in DNA. If one chromosome of a pair becomes irreversibly damaged, the information from the other chromosome can be copied and inserted by recombination to provide a correct replacement of the damaged section. The key idea here is that sequences flanking the damage from a sister chromosome can base-pair with the corresponding sequences on the damaged chromosome, thus allowing replication to copy the correct sequence and repair the lesion.

Site-specific Recombination

Site-specific recombination involves very short specific sequences that are recognized by proteins. Long DNA sequences such as viral genomes, drug-resistance elements, and regulatory sequences such as the mating type locus in yeast can be inserted, removed, or inverted, having profound regulatory effects. More than any other mechanism, site-specific recombination is responsible for reshaping genomes. For example, the genomes of many higher organisms, including plants and humans, show evidence that transposable elements have been constantly inserted throughout the genome and even into one another from time to time.

One example of site-specific recombination is the integration of DNA from bacteriophage λ into the chromosome of *E. coli*. In this reaction, bacteriophage λ DNA, which is a linear molecule in the normal phage, first forms a circle and then is cleaved by the enzyme λ-integrase at a specific site called the phage attachment site. A similar site on the bacterial chromosome is cut by integrase to give ends with the identical extension. Because of the complementarity between these two ends, they can be rejoined so that the original circular λ chromosome is inserted into the chromosome of the *E. coli* bacterium. Once integrated, the phage can be held in an inactive state until signals are generated that reverse the process, allowing the phage genome to escape and resume its normal life cycle of growth and spread into other bacteria. This site-specific recombination process requires only λ-integrase and one host DNA binding protein called the integration host factor. A third protein, called excisionase, recognizes the hybrid sites formed on integration and, in conjunction with integrase, catalyzes an excision process whereby the λ chromosome is removed from the bacterial chromosome.

A similar but more widespread version of DNA integration and excision is exhibited by the transposons, the so-called jumping genes. These elements range in size from fewer than 1,000 to as many as 40,000 base pairs. Transposons are able to move from one location in a genome to another, as first discovered in corn (maize) during the 1940s and '50s by Barbara McClintock, whose work won her a Nobel Prize in 1983. Most, if not all, transposons encode an enzyme called transposase that acts much like λ-integrase by cleaving the ends of the transposon as well as its target site.

Transposons differ from bacteriophage λ in that they do not have a separate existence outside of the chromosome but rather are always maintained in an integrated site. Two types of transposition can occur—one in which the element simply moves from one site in the chromosome to another and a second in which the transposon is replicated prior to moving. This second type of transposition leaves behind the original copy of the transposon and generates a second copy that is inserted elsewhere in the genome. Known as replicative transposition, this process is the mechanism responsible for the vast spread of transposable elements in many higher organisms.

The simplest kinds of transposons merely contain a copy of the transposase with no additional genes. They behave as parasitic elements and usually have no known associated function that is advantageous to the host. More often, transposable elements have additional genes associated with them—for example, antibiotic resistance factors. Antibiotic resistance typically occurs when an infecting bacterium acquires a plasmid that carries a gene encoding resistance to one or more antibiotics. Typically, these resistance genes are carried on transposable elements that have moved into plasmids and are easily transferred from one organism to another. Once a bacterium picks up such a gene, it enjoys a great selective advantage because it can grow in the presence of the antibiotic. Indiscriminate use of antibiotics actually promotes the buildup of these drug-resistant plasmids and strains.

Repair

It is extremely important that the integrity of DNA be maintained in order to ensure the accurate workings of a cell over its lifetime and to make certain that genetic information is accurately passed from one generation to the next. This maintenance is achieved by repair processes that constantly monitor the DNA for lesions and activate appropriate repair enzymes. Serious lesions in DNA such as pyrimidine dimers or gaps can be repaired by recombination mechanisms, but there are many other repair mechanisms.

One important mechanism is that of mismatch repair, which has been studied extensively in *E. coli*. The system is directed by the presence of a methyl group within the sequence GATC on the template strand. Comparable systems for mismatch repair also operate in eukaryotes, though the template strand is not marked by methyl groups. In fact, lesions within the genes for human mismatch repair systems are known to be responsible for many cancers. Loss of the mismatch repair system allows mutations to build up quickly and eventually to affect the genes that cause cells to divide. As a result, cells divide in an uncontrolled manner and become cancerous.

Once replication is complete, the most common kind of damage to nucleic acids is one in which the normal A, C, G, and T bases are changed into chemically modified bases that usually differ significantly from their natural counterparts. The only exceptions are the deamination of cytosine to uracil and the deamination of 5-methylcytosine to thymine. In these cases the product is a G:U or G:T mismatch. Specific enzymes called DNA glycosylases can recognize uracil in DNA or the thymine in a G:T mismatch and can selectively remove the base by cleaving the bond between the base and the deoxyribose sugar. Many of these enzymes are specific for the different chemically modified bases that may be present in DNA.

Another common means of repairing DNA lesions is by an excision repair pathway. Enzymes recognize damage within DNA, probably by detecting an altered conformation of DNA, and then nick the strand on either side of the lesion, allowing a small single-stranded DNA to be excised. DNA polymerase and DNA ligase then repair the single-stranded gap. In all of these systems, the presence of an abnormal base signifies which strand is to be repaired, and the complementary strand is used as the template to ensure the accuracy of repair.

RNA Metabolism

RNA provides the link between the genetic information encoded in DNA and the actual workings of the cell. Some RNA molecules such as the rRNAs and the snRNAs become part of complicated ribonucleoprotein structures with specialized roles in the cell. Others such as tRNAs play key roles in protein synthesis, while mRNAs direct the synthesis of proteins by the ribosome. Three distinct phases of RNA metabolism occur. First, selected segments of the genome are copied by transcription to produce the precursor RNAs. Second, these precursors are processed to become functionally mature RNAs ready for use.

When these RNAs are mRNAs, they are then used for translation. Third, after use the RNAs are degraded, and the bases are recycled. Thus, transcription is the process where a specific segment of DNA, a gene, is copied into a specific RNA that encodes a single protein or plays a structural or catalytic role. Translation is the decoding of the information within mRNA molecules that takes place on a specialized structure called a ribosome. There are important differences in both transcription and translation between prokaryotic and eukaryotic organisms.

Transcription

Small segments of DNA are transcribed into RNA by the enzyme RNA polymerase, which achieves this copying in a strictly controlled process. The first step is to recognize a specific sequence on DNA called a promoter that signifies the start of the gene. The two strands of DNA become separated at this point, and RNA polymerase begins copying from a specific point on one strand of the DNA using a ribonucleoside 5′-triphosphate to begin the growing chain. Additional ribonucleoside triphosphates are used as the substrate, and, by cleavage of their high-energy phosphate bond, ribonucleoside monophosphates are incorporated into the growing RNA chain. Each successive ribonucleotide is directed by the complementary base pairing rules of DNA.

Thus, a C in DNA directs the incorporation of a G into RNA, G is copied into C, T into A, and A into U. Synthesis continues until a termination signal is reached, at which point the RNA polymerase drops off the DNA, and the RNA molecule is released. In some cases this RNA molecule is the final mRNA. In other cases it is a pre-mRNA and requires further processing before it is ready for translation by the ribosome. Ahead of many genes in prokaryotes, there are signals called "operators" where specialized proteins called repressors bind to the DNA just upstream of the start point of transcription and prevent access to the DNA by RNA polymerase. These repressor proteins thus prevent transcription of the gene by physically blocking the action of the RNA polymerase. Typically, repressors are released from their blocking action when they receive signals from other molecules in the cell indicating that the gene needs to be expressed. Ahead of some prokaryotic genes are signals to which activator proteins bind that positively induce transcription.

Transcription in higher organisms is more complicated. First, the RNA polymerase of eukaryotes is a more complicated enzyme than the relatively simple five-subunit enzyme of prokaryotes. In addition, there are many more accessory factors that help to control the efficiency of the individual promoters. These accessory proteins are called transcription factors and typically respond to signals from within the cell that indicate whether transcription is required. In many human genes, several transcription factors may be needed before transcription can proceed efficiently. A transcription factor can cause either repression or activation of gene expression in eukaryotes.

During transcription, only one strand of the DNA is usually copied. This is called the template strand, and the RNA molecules produced are single-stranded. The DNA strand that would correspond to the mRNA is called the coding or sense strand, and it is not unusual for this to change from one gene to the next. In eukaryotes the initial product of transcription is called a pre-mRNA, which is extensively spliced before the mature mRNA is produced, ready for translation by the ribosome.

Translation

The process of translation uses the information present in the nucleotide sequence of mRNA to direct the synthesis of a specific protein for use by the cell. Translation takes place on the ribosomes—complex particles in the cell that contain RNA and protein. In prokaryotes the ribosomes are loaded onto the mRNA while transcription is still ongoing. Near the 5′ end of the mRNA, a short sequence of nucleotides signals the starting point for translation. It contains a few nucleotides called a ribosome binding site, or Shine-Dalgarno sequence. In E. coli the tetranucleotide GAGG is sufficient to serve as a binding site. This typically lies five to eight bases upstream of an initiation codon. The mRNA sequence is read three bases at a time from its 5′ end toward its 3′ end, and one amino acid is added to the growing chain from its respective aminoacyl tRNA, until the complete protein chain is assembled. Translation stops when the ribosome encounters a termination codon, normally UAG, UAA, or UGA. Special release factors associate with the ribosome in response to these codons, and the newly synthesized protein, tRNAs, and mRNA all dissociate. The ribosome then becomes available to interact with another mRNA molecule.

In eukaryotes the essence of protein synthesis is the same, although the ribosomes are more complicated. As with prokaryotic initiation, the signal sequence interacts with the 3′ end of the small subunit rRNA during formation of the initiation complex. The issue of fidelity is important during protein synthesis, but it is not as crucial as fidelity during replication. One mRNA molecule can be translated repeatedly to give many copies of the protein. When an occasional protein is mistranslated, it usually does not fold properly and is then degraded by the cellular machinery. However, proofreading mechanisms exist within the ribosome to ensure accurate pairing between the codon in the mRNA and the anticodon in the tRNA.

One of the crowning achievements of molecular biology was the elucidation during the 1960s of the genetic code. Principals in this effort were Har G. Khorana and Marshall W. Nirenberg, who shared a Nobel Prize in 1968. Khorana and Nirenberg used artificial templates and protein synthesizing systems in the test tube to determine the coding potential of all 64 possible triplet codons. The key feature of the genetic code is that the 20 amino acids are encoded by 61 codons. Thus, there is degeneracy in the code such that one amino acid is often specified by more than one codon. In the case of serine and leucine, six codons can be used for each.

Among organisms that have been examined in detail, the code appears to be almost universal, from bacteria through archaea to eukaryotes. The known exceptions are found in the mitochondria of humans and many other organisms as well as in some species of bacteria. The structure within the genetic code whereby many amino acids are uniquely coded by the first two bases of the codon strongly suggests that the code has itself evolved from a more primitive code involving 16 dinucleotides. How the individual amino acids became associated with the different codons remains a matter of speculation.

DNA-Triplte	RNA Triplte	Amino Acid	DNA-Triplte	RNA-Triplte	Amino Acid
AAA	UUU	Phenylalanine	ACA	UGU	Cysteine
AAG	UUC		ACG	UGC	
			ACC	UGG	Tryptophan
AAT	UUA	Leucine	ATA	UAU	Tyrosine
AAC	UUG		ATG	UAC	
GAA	CUU		ATA	UAA	(termination: end of specification)
GAG	CUA		ATC	UAG	
GAT	CUG		ACT	UGA	
AGA	UCU	Serine	GCA	CGU	Arginine
AGG	UCC		GCG	CGC	
AGT	UCA		GCT	CGA	
AGC	UCG		GCC	CCG	
TCA	AGU		TCT	AGA	
TCG	AGC		TCC	AGG	

GGA GGG GGT GGC	CCU CCC CCA CCG	Proline	GTA GTG	CAU CAC	Histidine
TAA TAG TAT	AUU AUC AUA	Isoleucine (I1eu)	GTT GTC	CAA CAG	Glutamine(GluN)
TAC	AUG	Methionine	GTC	AAU AAC	Asparagine(AspN)
TGA TGG TGT TGC	ACU ACC ACA ACG	Threonine	TTT TTC	AAA AAG	Lysine
CAA CAG CAT CAC	GUU GUC GUA GUG	Valine	CCA CCG CCT CCC	GGU GGC GGA GGG	Glysine
			CTA CTG	GAU GAU	Aspartic acid
CGA CGG CGT CGC	GCU GCC GCA GCG	Alanine	CTT CTC	GAA GAG	Glutamic acid

Nucleotide

A nucleotide is an organic molecule that is the building block of DNA and RNA. They also have functions related to cell signaling, metabolism, and enzyme reactions. A nucleotide is made up of three parts: a phosphate group, a 5-carbon sugar, and a nitrogenous base. The four nitrogenous bases in DNA are adenine, cytosine, guanine, and thymine. RNA contains uracil, instead of thymine. A nucleotide within a chain makes up the genetic material of all known living things. They also serve a number of functions outside of genetic information storage, as messengers and energy moving molecules. A series of three nucleotides within the DNA is known as a codon, and directs the proteins within the cell to attach a specific protein to a series specified by the rest of the DNA. Special codons even specify to the machinery where to stop and start the process. DNA translation, as it is known, converts the information from DNA into the language of proteins. This chain of amino acids can then be properly folded, and provide one of many functions within the cell.

Nucleotide Structure

Nucleotide structure is simple, but the structure they can form together is complex. Below is an image of DNA. This molecule consists of two strands which wrap around each other, forming hydrogen bonds in the middle of the structure for support. Each nucleotide within has a specific structure which enables this formation.

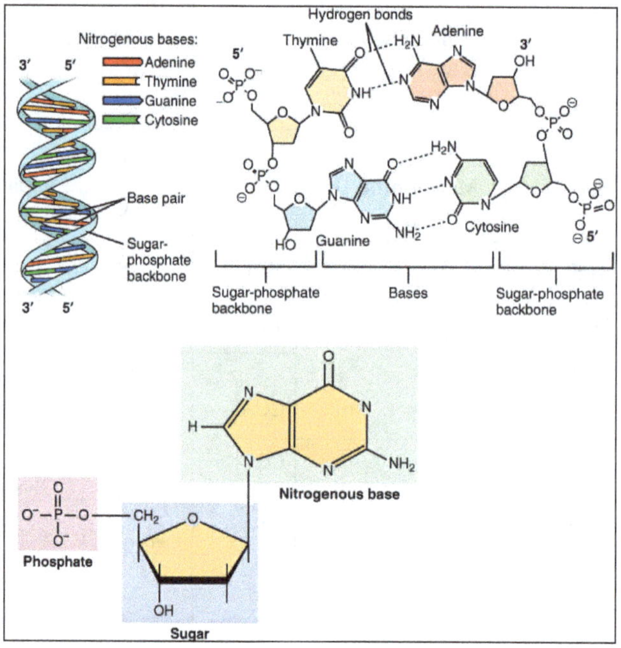

Nitrogenous Base

The nitrogenous base is the central information carrying part of the nucleotide structure. These molecules, which have different exposed functional groups, have differing abilities to interact with each other. As in the image, the idea arrangement is the maximum amount of hydrogen bonds between nucleotides involved. Because of the structure of the nucleotide, only a certain nucleotide can interact with other. The image above shows thymine bonding to adenine, and guanine bonding to cytosine. This is the proper and typical arrangement.

This even formation causes a twist in the structure, and is smooth if there are no errors. One of the ways proteins are able to repair damaged DNA is that they can bind to uneven spots within the structure. Uneven spots are created when hydrogen bonding does not occur between the opposing nucleotide molecules. The protein will cut out one nucleotide, and replace it with another. The duplicate nature of the genetic strands ensures that errors like this can be corrected with a high degree of accuracy.

Sugar

The second portion of the nucleotide is the sugar. Regardless of the nucleotide, the sugar is always the same. The difference is between DNA and RNA. In DNA, the 5-carbon sugar is deoxyribose, while in RNA, the 5-carbon sugar is ribose. This gives genetic molecules their names; the full name of DNA is deoxyribonucleic acid, and RNA is ribonucleic acid.

The sugar, with its exposed oxygen, can bond with the phosphate group of the next molecule. They then form a bond, which becomes the sugar-phosphate backbone. This structure adds rigidity to the structure, as the covalent bonds they form are much stronger than the hydrogen bonds between the two strands. When proteins come to process and transpose the DNA, they do so by separating the strands and reading only one side. When they pass on, the strands of genetic material comes back together, driven by the attraction between the opposing nucleotide bases. The sugar-phosphate backbone stays connected the whole time.

Phosphate Group

The last part of nucleotide structure, the phosphate group, is probably familiar from another important molecule ATP. Adenosine triphosphate, or ATP, is the energy molecule that most life on Earth relies upon to store and transfer energy between reactions. ATP contains three phosphate groups, which can store a lot of energy in their bonds. Unlike ATP, the bonds formed within a nucleotide are known as phosphodiester bonds, because they happen between the phosphate group and the sugar molecule.

During DNA replication, an enzyme known as DNA polymerase assembles the correct nucleotide bases, and begins organizing them against the chain it is reading. Another protein, DNA ligase, finished the job by creating the phosphodiester bond between the sugar molecule of one base and the phosphate group of the next. This creates the backbone of a new genetic molecule, able to be passed to the next generation. DNA and RNA contain all the genetic information necessary for cells to function.

Examples of Nucleotide

- Adenine: Adenine is a purine, which is one of two families of nitrogenous bases. Purines have a double-ringed structure. In DNA, adenine bonds with thymine. In RNA, adenine bonds with uracil. Adenosine triphosphate, uses the nucleotide adenine as a base. From there, three phosphate groups can be attached. This allows a great deal of energy to be stored in the bonds. For the same reason that the sugar-phosphate backbone is so strong, the bonds in ATP are as well. When combined with special enzymes which have formed to release the energy, it can be transferred to other reactions and molecules.
- Guanine: Like adenine, guanine is a purine nucleotide; it has a double ring. It bonds with cytosine in both DNA and RNA. As seen in the image above, guanine binds to cytosine through three hydrogen bonds. This makes the cytosine-guanine bond slightly stronger than the thymine-adenine bond, which only forms two hydrogen bonds.
- Cytosine: Pyrimidines are the other class of nucleotide. Cytosine is a pyrimidine nucleotide; it has only one ring in its structure. Cytosine bonds with guanine in both DNA and RNA. Bonding with the nucleotide guanine, the two make a strong pair.
- Thymine: Like the nucleotide cytosine, thymine is a pyrimidine nucleotide and has one ring. It bonds with adenine in DNA. Thymine is not found in RNA. In DNA, it forms only two hydrogen bonds with adenine, making them the weaker pair.
- Uracil: Uracil is also a pyrimidine. During transcription from DNA to RNA, uracil is placed

everywhere a thymine would normally go. The reason for this is not entirely understood, though uracil has some distinct advantages and disadvantages. Most creatures do not use uracil within the DNA because it is short lived, and can degrade into cytosine. However, in RNA uracil is the preferred nucleotide because RNA is also a short lived molecule.

Nucleotide Function

Besides being the basic unit of genetic material for all living things, a nucleotide can have other functions as well. A nucleotide can be a base in another molecule, such as adenosine triphosphate (ATP), which is the main energy molecule of the cell. They are also found in coenzymes like NAD and NADP, which come from ADP; these molecules are used in many chemical reactions that play roles in metabolism. Another molecule that contains a nucleotide is cyclic AMP (cAMP), a messenger molecule that is important in many processes including the regulation of metabolism and transporting chemical signals to cells. Nucleotides not only make up the building blocks of life, but also form many different molecules that function to make life possible.

Chromosome

Chromosomes are thread-like structures present in the nucleus, which carries genetic information from one generation to another. They play a vital role in cell division, heredity, variation, mutation, repair and regeneration. In Eukaryotic cells, genetic material is present in the nucleus in chromosomes, which is made up of highly organized DNA molecules with histone proteins supporting its structure. Chromosome means 'coloured body' that refers to its staining ability by certain dyes. Karl Nägeli, first observed the rod-like structure present in the nucleus of the plant cell. W. Waldeyer coined the term 'chromosome'. Walter Sutton and Theodor Boveri suggested that chromosomes are the physical carrier of genes in the eukaryotic cells.

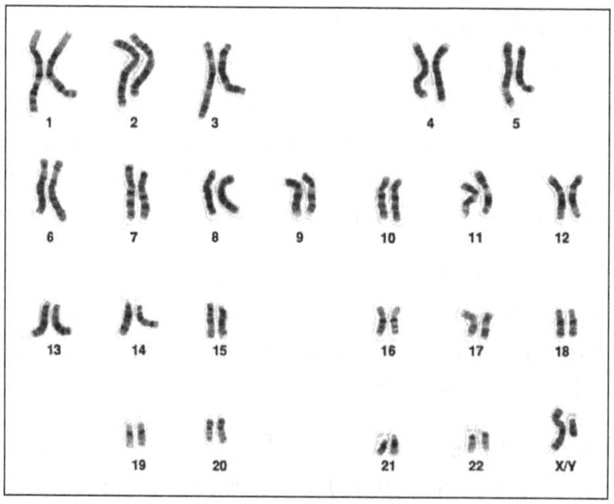

Karyotype of a human cell (male).

The number of chromosomes in any species is constant for all the cells. The number of chromosomes in gametes (e.g. sperms, egg) is half of the somatic cell and known as a haploid set of chromosomes, which is the result of meiosis during sexual reproduction. Chromosome number is preserved in the mitotic division of somatic cells, which is required for an organism to grow, repair and regenerate. Chromosome number varies in different species. A nematode species contains only 2 chromosomes in a cell, whereas a protozoan species contains as much as 1600 chromosomes in the cell. Most of the plant and animal species contain 8 to 50 number of chromosomes in its somatic cell. The number of chromosomes does not reflect the complexity of a species. A human cell contains total 23 pair of chromosomes (2n, total 23×2=46), of which 22 are autosomes and 1 sex chromosome. Karyotyping is a technique to study the structure of chromosomes present in a species. Chromosomes are isolated, stained and photographed. This technique is useful in finding out any chromosomal abnormalities.

Chromosome Structure

Each cell has a pair of each kind of chromosome known as a homologous chromosome. Chromosomes are made up of chromatin, which contains a single molecule of DNA and associated proteins. Each chromosome contains hundreds and thousands of genes that can precisely code for several proteins in the cell. Structure of a chromosome can be best seen during cell division.

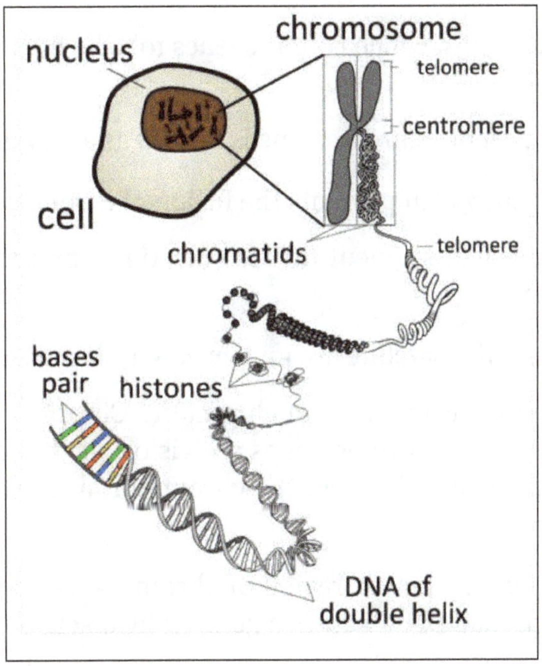

Main parts of chromosomes are:

- Chromatid: Each chromosome has two symmetrical structures called chromatids or sister chromatids which is visible in mitotic metaphase.
    - Each chromatid contains a single DNA molecule.
    - At the anaphase of mitotic cell division, sister chromatids separate and migrate to opposite poles.
- Centromere and kinetochore: Sister chromatids are joined by the centromere.
    - Spindle fibres during cell division are attached at the centromere.
    - The number and position of the centromere differs in different chromosomes.

  - The centromere is called primary constriction.
  - Centromere divides the chromosome into two parts, the shorter arm is known as 'p' arm and the longer arm is known as 'q' arm.
  - The centromere contains a disc-shaped kinetochore, which has specific DNA sequence with special proteins bound to them.
  - The kinetochore provides the centre for polymerisation of tubulin proteins and assembly of microtubules.
- Secondary constriction and nucleolar organisers: Other than centromere, chromosomes possess secondary constrictions.
  - Secondary constrictions can be identified from centromere at anaphase because there is bending only at the centromere (primary constriction).
  - Secondary constrictions, which contain genes to form nucleoli are known as the nucleolar organiser.
- Telomere: Terminal part of a chromosome is known as a telomere.
  - Telomeres are polar, which prevents the fusion of chromosomal segments.
- Satellite: It is an elongated segment that is sometimes present on a chromosome at the secondary constriction.
  - The chromosomes with satellite are known as sat-chromosome.
- Chromatin: Chromosome is made up of chromatin. Chromatin is made up of DNA, RNA and proteins. At interphase, chromosomes are visible as thin chromatin fibres present in the nucleoplasm. During cell division, the chromatin fibres condense and chromosomes are visible with distinct features:
  - The darkly stained, condensed region of chromatin is known as heterochromatin. It contains tightly packed DNA, which is genetically inactive.
  - The light stained, diffused region of chromatin is known as euchromatin. It contains genetically active and loosely packed DNA.
  - At prophase, the chromosomal material is visible as thin filaments known as chromonemata.
  - At interphase, bead-like structures are visible, which are an accumulation of chromatin material called chromomere. Chromatin with chromomere looks like a necklace with beads.

Structural Organisation of Chromatin

Chromatin consists of DNA and associated proteins. DNA is packaged in a highly organised manner in chromosomes.

- Nucleosomes are the basic unit of chromatin. It is 10 nm in the diameter.
  - DNA packing is facilitated by proteins called histones. DNA is wound around histone proteins to form a nucleosome.
  - There are 5 types of histone proteins in the eukaryotic chromosomes, namely H1, H2A, H2B, H3 and H4.
  - Histones are positively charged due to the presence of amino acids with basic side chains and it associates with negatively charged DNA due to the presence of phosphate groups.
  - Histone proteins play an important role in gene regulation.
  - Nucleosomes are made up of 146 base pair of DNA coiled around a core of eight histone molecules (2 molecules of 4 histone proteins).
  - Nucleosomes prevent DNA from getting tangled.
- Linker DNA and the fifth histone (H1) pack adjacent nucleosomes to a 30 nm compact chromatin fibre.
- These fibres form a large coiled loop held together by non-histone proteins (actin, α and β tubulin, myosin) called scaffolding proteins to form extended chromatin which is 300 nm in diameter.
- Chromatin further condenses with the help of protein known as condensin, it binds to DNA and wraps it into coiled loops and we get the compacted chromosome.

Types of Chromosome

Based on the number of centromeres present:

- Monocentric: Having only one centromere.
- Holocentric: Having diffused centromere and microtubules are attached along the length of a chromosome.
- Acentric: Chromosome may break and fuse together to form a chromosome without a centromere. It cannot attach to the mitotic spindle.
- Dicentric: Chromosomal aberration where chromosomes break and fuse together with two centromeres. They are also unstable as two centromeres tend to migrate to opposite poles resulting in fragmentation.

Based on the position of the centromere:

- Telocentric: Rod-like chromosome with centromere present on the proximal end. No 'p' arm (short arm) present. Telomeric chromosomes are not found in humans.
- Acrocentric: Rod-like, centromere present at one end giving rise to one very short arm and an exceptionally long arm.

- Submetacentric: L-shaped or J-shaped, with centromere near the centre of the chromosome giving rise to two unequal arms.
- Metacentric: V-shaped chromosomes with centromere in the middle giving rise to two equal arms.

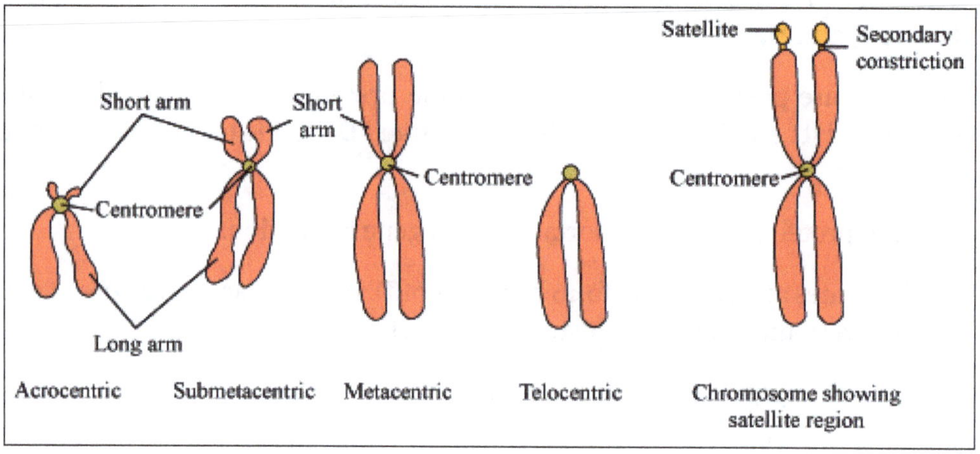

Giant Chromosomes

1. Polytene chromosome:

- Balbiani first discovered a structure in the nuclei of secretory glands of midges.
- Painter, Heitz and Bauer, rediscovered them in the salivary gland of Drosophila and recognised them as a chromosome.
- Also known as Salivary gland chromosome.
- These are called polytene by Kollar due to the presence of many chromonemata in them.
- These are present in some cells of the larvae of Dipteran insects.
- These are very large due to the presence of high DNA content.
- The polytene chromosome of Drosophila's salivary gland has 1000 DNA molecules Chironomus has 1600 DNA molecules in its each polytene chromosome.
- There is a series of alternating dark and clear bands called interband.
- Chromosome puffs or Balbiani rings are present, which are the swelling of bands due to DNA unfolding into open loops. These are the region of the intense transcription or mRNA formation.

2. Lampbrush chromosome:

- First discovered in the oocytes of salamander.
- The name is given due to its resemblance with a brush that is used for cleaning lamp, glass chimneys, etc.
- They occur at the diplotene stage of oocytes in vertebrates and invertebrates.

- Lampbrush chromosomes are also found in the spermatocytes of many animals and also found in the giant nucleus of algae Acetabularia.
- They are present as a bivalent with 4 chromatids.
- Chromosomal axis is formed from highly condensed chromatin and lateral loops extend from the row of chromomeres.
- Lateral loops of DNA are always symmetrical and formed due to intense RNA synthesis.
- In the oocytes of salamander, there are 10,000 loops present per haploid set of chromosomes.
- The centromere doesn't bear any loops.

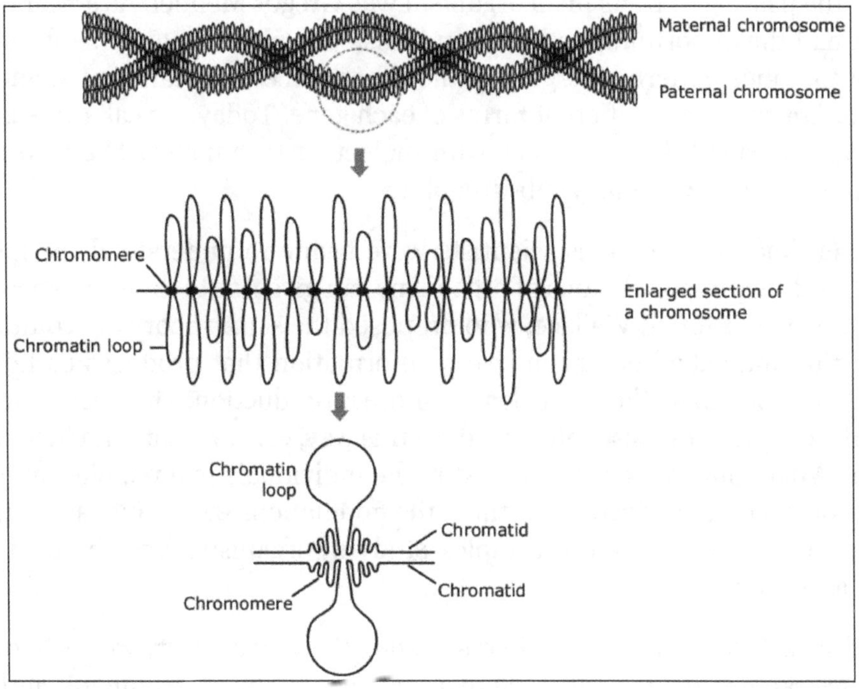

Functions of Chromosomes

- The main function of chromosomes is to carry the genetic material from one generation to another.
- Chromosomes play an important role and act as a guiding force in the growth, reproduction, repair and regeneration process that is important for their survival.
- Chromosomes protect the DNA from getting tangled and damaged.
- Histone and non-histone proteins help in the regulation of gene expression.
- Spindle fibres attached to centromere help in the movement of the chromosome during cell division.
- Each chromosome contains thousands of genes that precisely code for multiple proteins present in the body.

Heredity

Heredity is the passing of traits from parent to offspring. Molecules of DNA carry information that codes for various proteins. These proteins interact with the environment, causing observable patterns of life. The complex mechanisms that replicate and reproduce DNA and the organisms it creates can be recombined and mutated during the process, leading to new and various forms of life. All organisms, from the simplest bacteria to the largest eukaryotes, use DNA as the main form of heredity.

Before the role of DNA was understood, it was well known that some mechanism caused offspring to resemble the parents. Children look like their parents, livestock reproduce in predictable lines, and even plants have visible traits that pass from one generation to the next. The first scientist to fully document the passing of traits in an organism was Gregor Mendel. As a friar living in a monastery, Mendel had the opportunity to breed and raise pea plants, which he observed with great care. He began to notice pattern emerging in the inheritance of certain traits, and proposed the idea that each organism carries different forms of each gene. Today, we call these genetic variants alleles, and have confirmed their existence with molecular techniques. The field of genetics has grown into a large science, with many sub-disciplines.

Around the same time, other famous scientists were trying to understand the larger picture of heredity, and how different populations of organisms can give rise to different species. These men were Charles Darwin and Alfred Wallace, who proposed the same theory of evolution, separately. They proposed that individual organisms carry information that produces certain traits. Some traits are more beneficial than others, and lead to more reproduction. These traits are passed to the offspring, and the offspring can also interbreed. In this way, certain traits can increase or decrease in a population. When mutations or barriers stop the individuals in a population from reproducing, the population becomes divided. Over time, the populations evolve into separate species. The theory of evolution has evolved into a complex study of organisms and the environments they occupy, known as ecology.

Today, many of these fields interact, as scientists study the way heredity works in organisms. Molecular techniques can be used to analyze changes created by the environment, and natural selection acting on the alleles. Or, working the other way around, the genome can be altered to see what changes occur in the organism. Either ways, scientist now have a large arsenal of tools to analyze heredity, and are making serious advances in understanding the chemical and environmental forces that affect heredity. It is now even possible to change the DNA an organism inherits, and fix various mutations. As such, modern medicine has dedicated many resources to studying these mechanisms.

Examples of Heredity

Heredity in Bacteria

Bacteria are simple prokaryotic organisms. They are haploid in nature, and carry only one allele for each gene. Their genome is usually contained in a single chromosome, which exists in a ring. Bacteria reproduce through an asexual process known as binary fission. During binary fission, the DNA is copied, and the copies are segregated into new cells. The DNA in each cell exists in a double helix, one half of the helix being old DNA and the other half being newly copied DNA. In this way, each daughter bacteria is identical to the original parent.

This mode of heredity relies on mutations to change the alleles at each gene. When a mutation is beneficial, bacteria can reproduce more. If the environment changes and the allele is no longer beneficial, the population with the allele will suffer. Sometimes, these mutations can allow bacteria to survive certain antibiotics. Even this resistance to antibiotics is a heritable trait, and once the mutation happens in a population, it is hard to get rid of. If a population of harmful bacteria infect a human and antibiotics cannot get rid of them, the infection could become lethal. Scientists study modes of heredity in bacteria to develop new strategies to fight them in the field of public health.

Heredity in Sexually-reproducing Organisms

In sexually reproducing organisms, the mode of heredity gets more complicated. Instead of each individuals giving rise to their own offspring by simply copying the DNA, two organisms must combine their DNA to create offspring. This method is much more complex, but leads to more variation in the offspring, which can increase their chances of success in a changing world. Most sexually-reproducing organisms exists as diploids, with two alleles of each gene. In order to reproduce sexually, these organisms must produce haploid cells through the process of meiosis. Meiosis consists of two consecutive cellular divisions, in which the number of alleles is reduced to one per gene.

In some organisms, like humans, these haploid cells develop into gametes, which seek gametes of the opposite sex so fertilization can take place. Other organisms, such as ferns, have a separate life cycle as a haploid organism, which produces many gametes. In both systems, the parents pass traits on to the offspring in a complex, multiple-allele system. The interactions of these alleles can produce different phenotypes, which add to the variety seen.

Genetic Code

Genetic Code is the instructions in a gene that tell the cell how to make a specific protein. A, C, G, and T are the "letters" of the DNA code; they stand for the chemicals adenine (A), cytosine (C), guanine (G), and thymine (T), respectively, that make up the nucleotide bases of DNA. Each gene's code combines the four chemicals in various ways to spell out three-letter "words" that specify which amino acid is needed at every step in making a protein.

Overlapping vs. Nonoverlapping Codes

Figure shows the difference between an overlapping and a nonoverlapping code. In the example, a three-letter, or triplet, code is shown. For the nonoverlapping code, consecutive amino acids are specified by consecutive code words (codons), as shown at the bottom of figure. For an overlapping code, consecutive amino acids are encoded in the mRNA by codons that share some consecutive bases; for example, the last two bases of one codon may also be the first two bases of the next codon. Overlapping codons are shown in the upper part of figure. Thus, for the sequence AUUGCUCAG in a nonoverlapping code, the first three amino acids are encoded by the three triplets AUU, GCU, and CAG, respectively. However, in an overlapping code, the first three amino acids are encoded by the triplets AUU, UUG, and UGC if the overlap is two bases, as shown in figure.

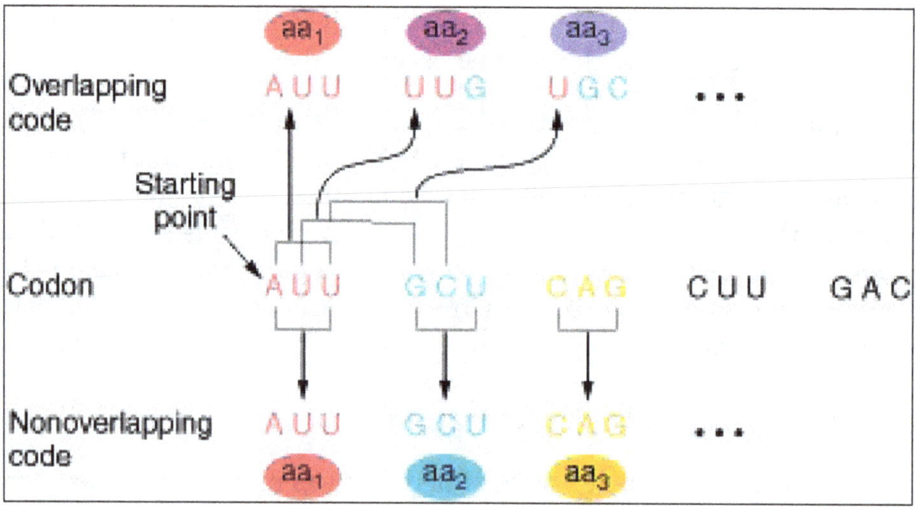

The difference between an overlapping and a nonoverlapping code. The case illustrated is for a code with three letters (a triplet code). An overlapping code uses codons that employ some of the same nucleotides as those of other codons for the translation of a single protein, as shown at the top of the diagram (for the mRNA sequence shown at the bottom of the diagram). In a nonoverlapping code, a protein is translated by reading codons that do not share any of the same nucleotides. Note that the designation of an amino acid as aa_2, aa_3, and so forth, in the nonoverlapping model does not mean it is necessarily the same amino acid as its numerical counterpart in the overlapping model. The reason is that the triplets making up the respective codons for the two aa_3's, for example, are different and would more than likely encode different amino acids. Identical amino acid numbers between the two models merely indicate the same amino acid position on the protein chain.

By 1961, it was already clear that the genetic code was nonoverlapping. The analysis of mutationally altered proteins, in particular, the nitrous acid–generated mutants of tobacco mosaic virus, showed that only a single amino acid changes at one time in one region of the protein. This result is predicted by a nonoverlapping code. As you can see from figure, an overlapping code predicts that a single base change will alter as many as three amino acids at adjacent positions in the protein.

It should be noted that, although the use of an overlapping *code* was ruled out by the analysis of single proteins, nothing precluded the use of alternative reading frames to encode amino acids in two different proteins. In the example here, one protein might be encoded by the series of codons that reads AUU, GCU, CAG, CUU, and so forth. A second protein might be encoded by codons that are shifted over by one base and therefore read UUG, CUC, AGC, UUG, and so forth. This is an example of storing the information encoding two different proteins in two different reading frames, while still using a genetic code that is read in a nonoverlapping manner during the translation of a *specific* protein. Some examples of such shifts in reading frame have been found.

Number of Letters in the Code

In reading an mRNA molecule from one particular end, only one of four different bases, A, U, G, or C, can be found at each position. Thus, if the words were one letter long, only four words would be possible. This vocabulary cannot be the genetic code, because we must have a word for each of the 20 amino acids commonly found in cellular proteins. If the words were two letters long, then $4^2 = 16$ words would be possible; for example, AU, CU, or CC. This vocabulary is still not large enough.

If the words are three letters long, then $4^3 = 64$ words are possible; for example, AUU, GCG, or UGC. This vocabulary provides more than enough words to describe the amino acids. We can conclude that the code word must consist of at least three nucleotide pairs. However, if all words are "triplets," then we have a considerable excess of possible words over the 20 needed to name the common amino acids.

Use of Suppressors to Demonstrate a Triplet Code

Convincing proof that a codon is, in fact, three letters long (and no more than three) came from beautiful genetic experiments first reported in 1961 by Francis Crick, Sidney Brenner, and their co-workers, who used mutants in the *rII* locus of T4 phage. Mutations causing the rII phenotype were induced by using a chemical called *proflavin,* which was thought to act by the addition or deletion of single nucleotide pairs in DNA. (This assumption is based on experimental evidence not presented here.) The following examples illustrate the action of proflavin on double-stranded DNA.

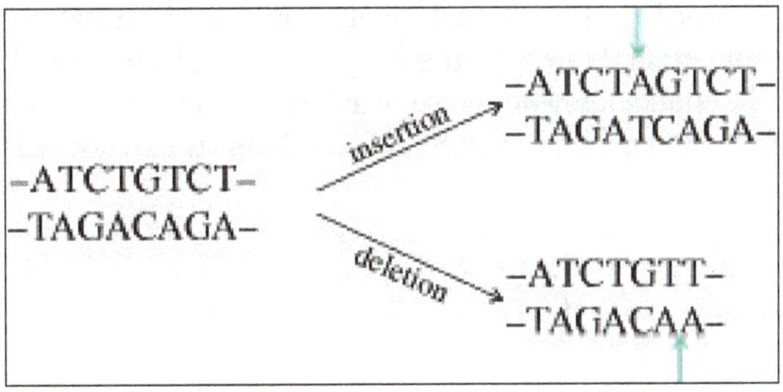

Then, starting with one particular proflavin-induced mutation called FCO, Crick and his colleagues found "reversions" (reversals of the mutation) that were detected by their wild-type plaques on *E. coli* strain K(λ). Genetic analysis of these plaques revealed that the "revertants" were not identical true wild types, thereby suggesting that the back mutation was not an exact reversal of the original forward mutation. In fact, the reversion was found to be caused by the presence of a *second mutation* at a different site from—but in the same gene as—that of FCO; this second mutation "suppressed" mutant expression of the original FCO. a suppressor mutation counteracts or suppresses the effects of another mutation. The suppressor mutation could be separated from the original forward mutation by recombination, and, as we have seen, when this was done, the suppressor was shown to be an *rII* mutation itself.

The suppressor of an initial *rII* mutation is shown to be an *rII* mutation itself after separation by crossing over. The original mutant, FCO, was induced by proflavin. Later, when the FCO strain was treated with proflavin again, a revertant was found, which on first appearance seemed to be wild type. However, a second mutation within the *rII* region was found to have been induced, and the double mutant $rII_x rII_y$ was shown not to be quite identical with the original wild type.

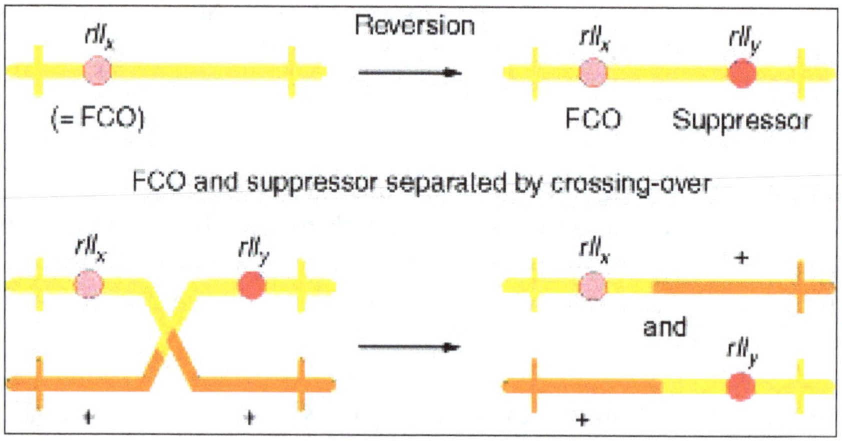

How can we explain these results? If we assume that reading is polarized—that is, if the gene is read from one end only—then the original proflavin-induced addition or deletion could be mutant because it interrupts a normal reading mechanism that establishes the group of bases to be read as words. For example, if each three bases on the resulting mRNA make a word, then the "reading frame" might be established by taking the first three bases from the end as the first word, the next three as the second word, and so forth. In that case, a proflavin-induced addition or deletion of a single pair on the DNA would shift the reading frame on the mRNA from that corresponding point on, causing all following words to be misread. Such a frameshift mutation could reduce most of the genetic message to gibberish. However, the proper reading frame could be restored by a compensatory insertion or deletion somewhere else, leaving only a short stretch of gibberish between the two. Consider the following example in which three-letter English words are used to represent the codons:

	THE FAT CAT ATE THE BIG RAT
Delete C:	THE FAT ATA TET HEB IGR AT ↑
Insert A:	THE FAT ATA ATE THE BIG RAT ↑

The insertion suppresses the effect of the deletion by restoring most of the sense of the sentence. By itself, however, the insertion also disrupts the sentence:

THE FAT CAT AAT ETH EBI GRA T

If we assume that the FCO mutant is caused by an addition, then the second (suppressor) mutant would have to be a deletion because, as we have seen, this would restore the reading frame of the resulting message (a second insertion would not correct the frame). In the following diagrams, we use a hypothetical nucleotide chain to represent RNA for simplicity. We also assume that the code words are three letters long and are read in one direction.

1. Wild-type message:

CAU CAU CAU CAU CAU

2. *rII_a* message: Distal words changed (x) by frameshift mutation (words marked ✓ are unaffected).

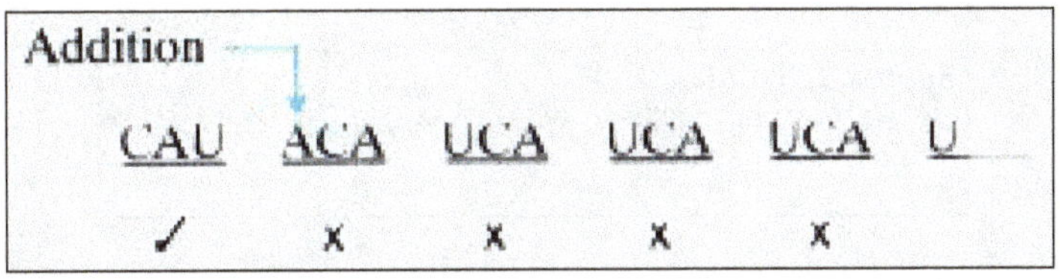

3. *rII_a rII_b* message: Few words wrong, but reading frame restored for later words.

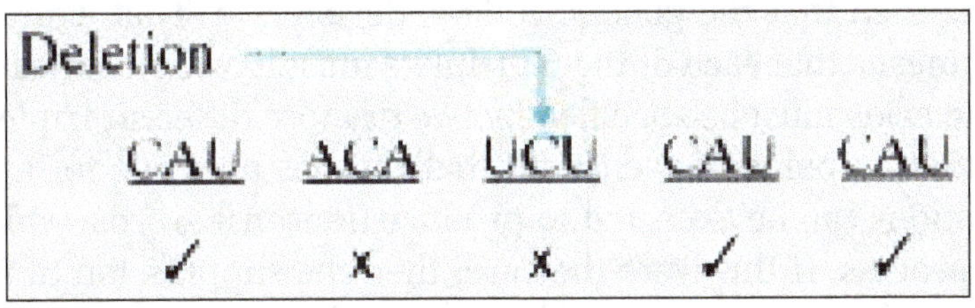

The few wrong words in the suppressed genotype could account for the fact that the "revertants" (suppressed phenotypes) that Crick and his associates recovered did not look exactly like the true wild types phenotypically.

We have assumed here that the original frameshift mutation was an addition, but the explanation works just as well if we assume that the original FCO mutation is a deletion and the suppressor is an addition. If the FCO is defined as plus, then suppressor mutations are automatically minus. Experiments have confirmed that a plus cannot suppress a plus and a minus cannot suppress a minus. In other words, two mutations of the same sign never act as suppressors of each other. However, very interestingly, combinations of three pluses or three minuses have been shown to act together to restore a wild-type phenotype. This observation provided the first experimental confirmation that a word in the genetic code consists of three successive nucleotide pairs, or a triplet. The reason is that three additions or three deletions within a gene automatically restore the reading frame in the mRNA if the words are triplets. For example:

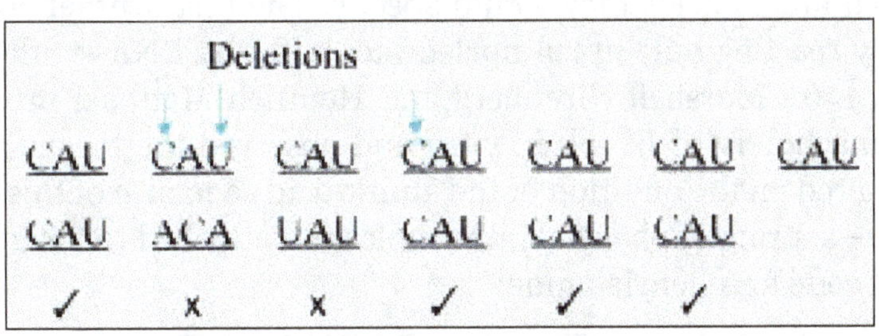

Proof that the genetic deductions about proflavin were correct came from an analysis of proflavin-induced mutations in a gene with a protein product that could be analyzed. George Streisinger worked with the gene that controls the enzyme lysozyme, which has a known amino acid sequence. He induced a mutation in the gene with proflavin and selected for proflavin-induced revertants, which were shown genetically to be double mutants (with mutations of opposite sign). When the protein of the double mutant was analyzed, a stretch of different amino acids lay between two wild-type ends, just as predicted.

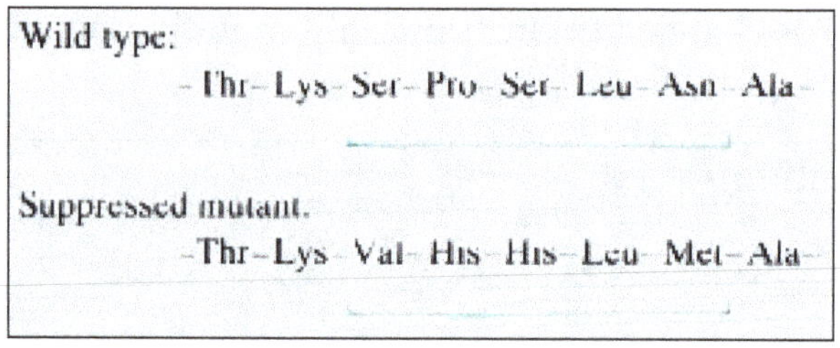

Degeneracy of the Genetic Code

Crick's work also suggested that the genetic code is degenerate. That expression is not a moral indictment. It simply means that each of the 64 triplets must have some meaning within the code; so at least some amino acids must be specified by two or more different triplets. If only 20 triplets are used (with the other 44 being nonsense, in that they do not code for any amino acid), then most frameshift mutations can be expected to produce nonsense words, which presumably stops the protein-building process. If this were the case, then the suppression of frameshift mutations would rarely, if ever, work. However, if all triplets specified some amino acid, then the changed words would simply result in the insertion of incorrect amino acids into the protein. Thus, Crick reasoned that many or all amino acids must have several different names in the base-pair code; this hypothesis was later confirmed biochemically.

Cracking the Code

The deciphering of the genetic code—determining the amino acid specified by each triplet—was one of the most exciting genetic breakthroughs of the past 50 years. Once the necessary experimental techniques became available, the genetic code was broken in a rush. The first breakthrough was the discovery of how to make synthetic mRNA. If the nucleotides of RNA are mixed with a special enzyme (polynucleotide phosphorylase), a single-stranded RNA is formed in the reaction. No DNA is needed for this synthesis, and so the nucleotides are incorporated at random.

The ability to synthesize mRNA offered the exciting prospect of creating specific mRNA sequences and then seeing which amino acids they would specify. The first synthetic messenger obtained, poly(U), was made by reacting only uracil nucleotides with the RNA-synthesizing enzyme, producing –UUUU–. In 1961, Marshall Nirenberg and Heinrich Matthaei mixed poly(U) with the proteinsynthesizing machinery of *E. coli* in vitro and observed the formation of a protein. The main excitement centered on the question of the amino acid sequence of this protein. It proved to be polyphenylalanine—a string of phenylalanine molecules attached to form a polypeptide. Thus, the triplet UUU must code for phenylalanine:

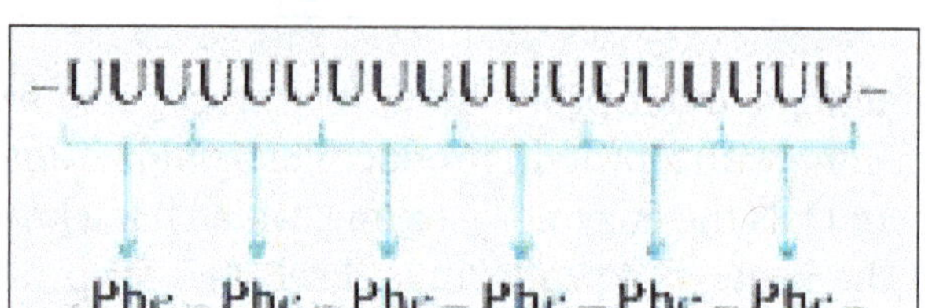

This type of analysis was extended by mixing nucleotides in a known fixed proportion when making synthetic mRNA. In one experiment, the nucleotides uracil and guanine were mixed in a ratio of 3:1. When nucleotides are incorporated at random into synthetic mRNA, the relative frequency at which each triplet will appear in the sequence can be calculated on the basis of the relative proportion of the various nucleotides present. Note that, in table, UUU is used as the baseline frequency against which the other frequencies are measured in determining their respective ratios. For example, UUG, with a probability of *p*(UUG) = 9/64, would be expected only one-third as often as UUU, with its probability of *p*(UUU) = 27/64. Stated alternatively, *p*(UUG)/*p*(UUU) = 9/27 = 1/3 = 0.33, which is the ratio for UUG given in table.

Table: Expected Frequencies of Various Codons in Synthetic mRNA Composed of 3/4 Uracil and 1/4 Guanine.

Codon	Probability	Ratio*
UUU	p(UUU) =3/4 × 3/4 × 3/4 =27/64	1.00
UUG	p(UUG) =3/4 × 3/4 × 1/4 =9/64	0.33
UGU	p(UGU) =3/4 × 1/4 × 3/4 =9/64	0.33
GUU	p(GUU) =1/4 × 3/4 × 3/4 =9/64	0.33
UGG	p(UGG) =3/4 × 1/4 × 1/4 =3/64	0.11
GGU	p(GGU) =1/4 × 1/4 × 3/4 =3/64	0.11
GUG	p(GUG) =1/4 × 3/4 × 1/4 =3/64	0.11
GGG	p(GGG) =1/4 × 1/4 × 1/4 =1/64	0.03

*UUU is used as the baseline frequency against which the frequencies of the other codons are measures in establishing the respective ratios. For example, the ratio for UUG is derived from *p*(UUG)/*p*(UUU) = 0.33.

If these codons each encode a different amino acid (that is, are not redundant), we expect the amino acids generated by this particular mix of guanine and uracil to be in ratios similar to those of the various codons. Although there is some redundancy among these codons, the ratios of the amino acids actually obtained from this mix of bases are indeed quite similar to the ratios seen for the codon frequencies in table.

Table: Observed Frequencies of Various Amino Acids in Protein Translated from mRNA Composed of 3/4 Uracil and 1/4 Guanine.

Amino acid	Ratio
Phenylalanine	1.00
Leucine	0.37
Valine	0.36
Cysteine	0.35
Tryptophan	0.14
Glycine	0.12

From this evidence, we can deduce that codons consisting of one guanine and two uracils (G + 2 U) code for valine, leucine, and cysteine, although we cannot distinguish the specific sequence for each of these amino acids. Similarly, one uracil and two guanines (U + 2 G) must code for tryptophan, glycine, and perhaps one other. It looks as though the Watson-Crick model is correct in predicting the importance of the precise sequence (not just the ratios of bases). Many provisional assignments (such as those just outlined for G and U) were soon obtained, primarily by groups working with Nirenberg or with Severo Ochoa.

tRNA Recognition of the Codon

Is it the tRNA or the amino acid itself that recognizes the mRNA that encodes a specific amino acid? A very convincing experiment answered this question. In the experiment, an aminoacyl-tRNA (aa-tRNA), cysteinyl-tRNA ($tRNA^{Cys}$, the tRNA specific for cysteine) "charged" with cysteine was treated with nickel hydride, which converted the cysteine (while still bound to $tRNA^{Cys}$) into another amino acid, alanine, without affecting the tRNA:

$$\text{cysteine} - \text{tRNA}^{Cys} \xrightarrow{\text{nickel hydride}} \text{alanine} - \text{tRNA}^{Cys}$$

Protein synthesized with this hybrid species had alanine wherever we would expect cysteine. Thus, the experiment demonstrated that the amino acids are "illiterate"; they are inserted at the proper position because the tRNA "adapters" recognize the mRNA codons and insert their attached amino acids appropriately. We would expect, then, to find some site on the tRNA that recognizes the mRNA codon by complementary base pairing.

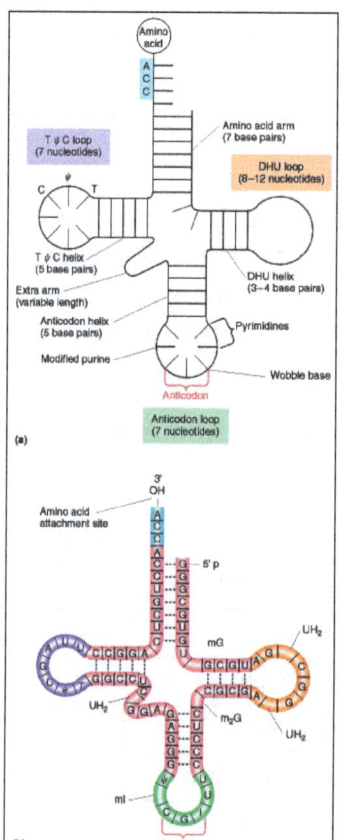

The structure of transfer RNA: (a) The functional areas of a generalized tRNA molecule. (b) The specific sequence of yeast alanine tRNA. Arrows indicate several kinds of rare modified bases.

(c) Diagram of the actual three-dimensional structure of yeast phenylalanine tRNA. The abbreviations ψ, mG, m_2G, mI, and DHU (or UH_2) are abbreviations for modified bases pseudouridine, methylguanosine, dimethylguanosine, methylinosine, and dihydrouridine, respectively.

Figure shows several functional sites of the tRNA molecule. The site that recognizes an mRNA codon is called the anticodon; its bases are complementary and antiparallel to the bases of the codon. Another operationally identifiable site is the amino acid attachment site. The other arms probably assist in binding the tRNA to the ribosome. Figure shows a specific tRNA (yeast alanine tRNA). The "flattened" cloverleafs shown in these diagrams are not the normal conformation of tRNA molecules; tRNA normally exists as an L-shaped folded cloverleaf, as shown in figure. These diagrams are supported by very sophisticated chemical analysis of tRNA nucleotide sequences and by X-ray crystallographic data on the overall shape of the molecule. Although tRNA molecules have many structural similarities, each has a unique three-dimensional shape that allows recognition by the correct synthetase, which catalyzes the joining of a tRNA with its specific amino acid to form an aminoacyl-tRNA. The specificity of charging the tRNAs is crucial to the integrity of protein synthesis.

Where does tRNA come from? If radioactive tRNA is put into a cell nucleus in which the DNA has been partly denatured by heating, the radioactivity appears (by autoradiography) in localized regions of the chromosomes. These regions probably indicate the location of genes that specify tRNA; they are regions of DNA that produce tRNA rather than mRNA, which produces a protein. The labeled tRNA hybridizes to these sites because of the complementarity of base sequences between the tRNA and its parent gene. A similar situation holds for rRNA. Thus, we see that even the one-gene–one-polypeptide idea is not completely valid. Some genes do not code for protein; rather, they specify RNA components of the translational apparatus.

How does tRNA get its fancy shape? It probably folds up spontaneously into a conformation that produces maximal stability. Transfer RNA contains many "odd" or modified bases (such as pseudouracil, ψ) in its nucleotides; these bases play a direct role in folding and have been implicated in other tRNA functions. You may have noticed some unusual base pairing within the loops of the tRNA in figure; G is hydrogen bonded to U (instead of C).

The Complete Code

Specific code words were finally deciphered through two kinds of experiments. The first required making "mini mRNAs," each only three nucleotides in length. These mini mRNAs are too short to promote translation into protein, but they do stimulate the binding of aminoacyl-tRNAs to ribosomes in a kind of abortive attempt at translation. It is possible to make a specific mini mRNA and determine *which* aminoacyl-tRNA that it will bind to ribosomes. For example, the G + 2 U problem can be resolved by using the following mini mRNAs:

- *GUU* stimulates binding of valyl-tRNA.
- *UUG* stimulates binding of leucyl -tRNA.
- *UGU* stimulates binding of cyesteiyl –tRNA.

[illegible] mini RNAs provided 64 possible codons. The second kind of experiment that was useful in cracking the genetic code required the use of repeating copolymers. For instance, the copolymer designated $(AGA)_n$, which is a long sequence of AGAAGAAGAAGAAGA, was used to stimulate polypeptide synthesis in vitro. From the sequence of the resulting polypeptides and the possible triplets that could reside in the respective RNA copolymer, many code words could be verified. Figure gives the genetic code dictionary of 64 words. Inspect this dictionary carefully, and ponder the miracle of molecular genetics. Such an inspection should reveal several points that require further explanation.

First letter	Second letter: U	Second letter: C	Second letter: A	Second letter: G	Third letter
U	UUU, UUC Phe UUA, UUG Leu	UCU, UCC, UCA, UCG Ser	UAU, UAC Tyr UAA Stop UAG Stop	UGU, UGC Cys UGA Stop UGG Trp	U C A G
C	CUU, CUC, CUA, CUG Leu	CCU, CCC, CCA, CCG Pro	CAU, CAC His CAA, CAG Gln	CGU, CGC, CGA, CGG Arg	U C A G
A	AUU, AUC, AUA Ile AUG Met	ACU, ACC, ACA, ACG Thr	AAU, AAC Asn AAA, AAG Lys	AGU, AGC Ser AGA, AGG Arg	U C A G
G	GUU, GUC, GUA, GUG Val	GCU, GCC, GCA, GCG Ala	GAU, GAC Asp GAA, GAG Glu	GGU, GGC, GGA, GGG Gly	U C A G

The genetic code.

Multiple Codons for a Single Amino Acid

We know that the number of codons for a single amino acid varies, ranging from one (tryptophan = UGG) to as many as six (serine = UCU or UCC or UCA or UCG or AGU or AGC). Why? The answer is complex but not difficult; it can be divided into two parts:

- Certain amino acids can be brought to the ribosome by several alternative tRNA types (species) having different anticodons, whereas certain other amino acids are brought to the ribosome by only one tRNA.
- Certain tRNA species can bring their specific amino acids in response to several codons, not just one, through a loose kind of base pairing at one end of the codon and anticodon. This sloppy pairing is called wobble.

We had better consider wobble first, and it will lead us into a discussion of the various species of tRNA. Wobble is caused by the third nucleotide of an anticodon (at the 5′ end) that is not quite aligned. This out-of-line nucleotide can sometimes form hydrogen bonds not only with its normal complementary nucleotide in the third position of the codon, but also with a different nucleotide in that position. Crick established certain "wobble rules" that dictate which nucleotides can and cannot form new hydrogen-bonded associations through wobble. In table, I (inosine) is one of the rare bases found in tRNA, often in the anticodon.

In the third site (5′ end) of the anticodon, G can take either of two wobble positions, thus being able to pair with either U or C. This ability means that a single tRNA species carrying an amino acid (in this case, serine) can recognize two codons—UCU and UCC—in the mRNA.

Table: Codon-anticodon pairings allowed by the wobble rules.

5′ end of anticodon	3′ end of codon
G	U or C
C	G only
A	U only
U	A or G
I	U, C, or A

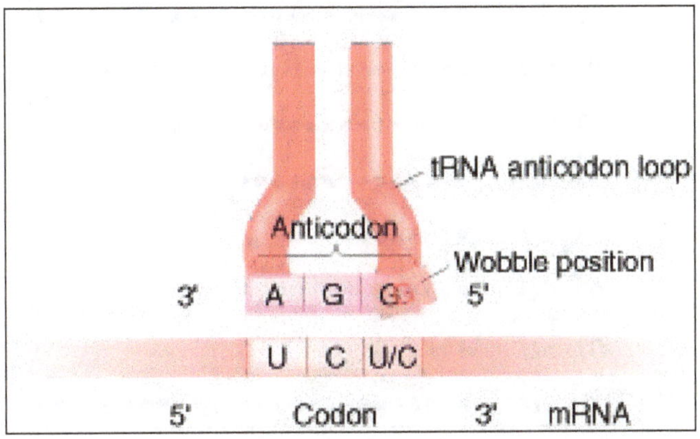

Figure shows the possible codons that one tRNA serine species can recognize. As the wobble rules indicate, G can pair with U or with C. Table lists all the codons for serine and shows how different tRNAs can service these codons. Serine affords a good example of the effects of wobble on the genetic code.

Table: Different tRNAs that can service codons for serine.

Codon	tRNA	Anticodon
UCU		
UCC	$tRNA^{Ser1}$	AGG + wobble
UCA		
UCG	$tRNA^{Ser2}$	AGU + wobble
AGU		
AGC	$tRNA^{Ser3}$	UCG + wobble

Sometimes there can be an additional tRNA species that we represent as $tRNA^{Ser4}$; it has an anticodon identical with any of the three anticodons shown in table 10-6, but it differs in its nucleotide sequence elsewhere in the tRNA molecule. These four tRNAs are called isoaccepting tRNAsbecause they accept the same amino acid, but they are probably all transcribed from different tRNA genes.

Stop Codons

The second point that you may have noticed in figure is that some codons do not specify an amino acid at all. These codons are labeled as stop or termination codons. They can be regarded as being similar to periods or commas punctuating the message encoded in the DNA. One of the first indications of the existence of stop codons came in 1965 from Brenner's work with the T4 phage. Brenner analyzed certain mutations ($m_1 - m_6$) in a single gene that controls the head protein of the phage. These mutants had two things in common. First, the head protein of each mutant was a shorter polypeptide chain than that of the wild type. Second, the presence of a suppressor mutation (su) in the host chromosome would cause the phage to develop a head protein of normal (wild-type) chain length despite the presence of the *m* mutation.

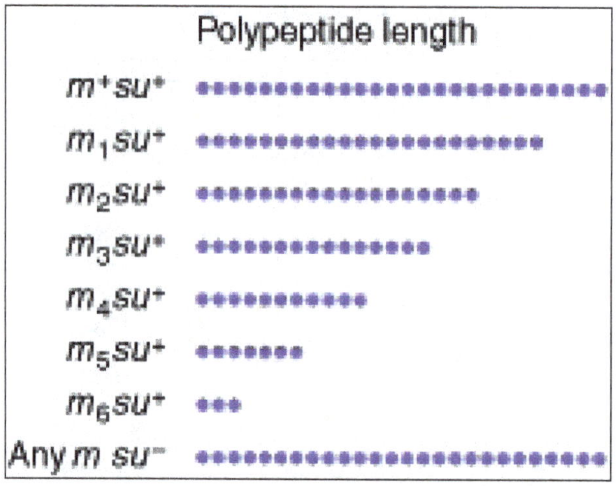

Polypeptide chain lengths of phage T4 head protein in wild type (top) and various amber mutants (m). An amber suppressor (su) leads to phenotypic development of the wild-type chain.

Brenner examined the ends of the shortened proteins and compared them with wild-type protein, recording for each mutant the next amino acid that *would* have been inserted to continue the wild-type chain. These amino acids for the six mutations were glutamine, lysine, glutamic acid, tyrosine, tryptophan, and serine. There is no immediately obvious pattern to these results, but Brenner brilliantly deduced that certain codons for each of these amino acids are similar in that each of them can mutate to the codon UAG by a single change in a DNA nucleotide pair. He therefore postulated that UAG is a stop (termination) codon—a signal to the translation mechanism that the protein is now complete.

UAG was the first stop codon deciphered; it is called the amber codon. Mutants that are defective owing to the presence of an abnormal amber codon are called amber mutants, and their suppressors are amber suppressors. UGA, the opal codon, and UAA, the ochre codon, also are stop codons and also have suppressors. Stop codons are often called nonsense codons because they designate no amino acid. Not surprisingly, stop codons do not act as mini mRNAs in binding aa-tRNA to ribosomes in vitro. We shall consider stop codons and their suppressors further after we have dealt with the process of protein synthesis.

Genotype

The genotype of an organism is the chemical composition of its DNA, which gives rise to the phenotype, or observable traits of an organism. A genotype consists of all the nucleic acids present in a DNA molecule that code for a particular trait. The outward appearance, or phenotype, is the result of interactions of proteins being created by the DNA. Modern DNA analyzing techniques have made it easier to identify which segments of DNA are responsible for various phenotypes.

A genotype has different alleles, or forms. The different alleles are produced by mutations to the DNA, and may give rise to beneficial or detrimental changes. In bacteria, the DNA exists in a ring and only one allele for each genotype is present. Sometime, an allele will mutate in a beneficial way, the organism will reproduce more and the genotype will increase in the population. In sexually reproducing organisms, there are two alleles present in each organism, which can have complex interactions with each other, and other genes. Mutations can occur in these alleles, new combinations can arise during meiosis, and infinite amount of variety can be created. These combinations of genotype give rise to the enormous variety of life on earth.

Examples of Genotype

Eye Color

Although a particular genotype consists of many nucleic acids, scientists typically represent genotypes with single letters, or two letters in the case of sexually reproducing organisms that receive one allele from each parent. For example, the trait for eye color could be represented with the letter "E". Varieties, or alleles, of that trait that are dominant will be designated by capital letters. Therefore, "E" will represent brown eyes. Traits that are recessive are written in lower case. The allele for blue eyes is recessive to the allele for brown eyes, so we can call it "e".

A set of parents has brown eyes. Having brown eyes only tells us their phenotype, not their genotype. The parents could be "Ee", "EE", or there could be one of both. A single "E" allele in the genotype will result in the brown-eyed phenotype, even if the parent harbors a recessive "e" allele as well. The parents conceive a child. The child has blue eye. This tells us that the child is homozygous recessive, or "ee", because only two recessive alleles can produce blue eyes. This also tells us a lot about the parents. The baby, having two "e" alleles, got one from each parent. Therefore, each parent has one "e" allele to give, while having the brown eyed phenotype. This shows us that the parents have the heterozygous "Ee" genotype. If either parent where homozygous dominant, "EE", the baby would have received at least one dominant "E" allele, giving it brown eyes.

Cystic Fibrosis

For a long time, it was not known why some children would develop a thick mucus in their airways, causing them to be short of breath and wheezy. The children had a variety of other symptoms, such as the inability to process food efficiently, gas, and weight loss. Until recent advances in medical and genetic sciences, many children died at a very early age. After years of research, cystic fibrosis was found to be caused by a defect in a gene that produces salt channels across cell membranes. These salt, or ion-channels, are used to maintain pH levels in various cells, remove waste, and remove nutrients from the intestines. The genotype of people with cystic fibrosis is homozygous recessive. In other words, they carry two copies of the non-functioning allele for the gene that creates specific ion-channels.

Some people, known as "carriers" can have a functioning, normal phenotype, while having a heterozygous genotype. This means that a carrier can pass a non-functioning allele on to their child. Unknowingly, two carriers can both pass on the non-functioning allele, and the child with receive a non-functioning homozygous recessive genotype. However, if one or both carrier parents passes on their good allele, the child will not have the symptoms of cystic fibrosis. If the child receives one functioning and one non-functioning allele, they will be a carrier as well. The child will not be carrier if they receive two functional alleles.

However, in parents who are both carriers, the genotypic ratio of offspring will be set at 1 normal: 2 carriers: 1 cystic fibrosis genotype. This genotypic ratio can be counted in a Punnett square. Place the heterozygous parents on the sides of the square, separating their individual alleles (Aa and Aa below). Then, simply fill in the boxes with the two alleles that would be received by each potential child. It can quickly be seen by counting that the genotypic ratio is 1AA:2Aa:1aa. In this case the phenotypic ratio would be 3 normal: 1 cystic fibrosis.

Phenotype

Phenotype refers to the observable characteristics of an organism as a multifactorial consequence of genetic traits and environmental influences. The organism phenotype includes its morphological, biochemical, physiological, and behavioral properties. The phenotype, therefore, is the total characteristics displayed by an organism that results from the expression of the genes of an organism as well as the influence of environmental factors and random variation. The interaction between these factors has often been represented by the following relationship: genotype + environment + random variation → phenotype.

In biology, the term "phenotype" is defined as the observable and measurable characteristics of an organism as a result of the interaction of the genes of the organism, environmental factors, and random variation. The phenotype of an organism will not only entail the observable features such as morphology but it will also include molecules and structures such as RNA and proteins produced as coded by the genes; this is referred to as "molecular phenotype".

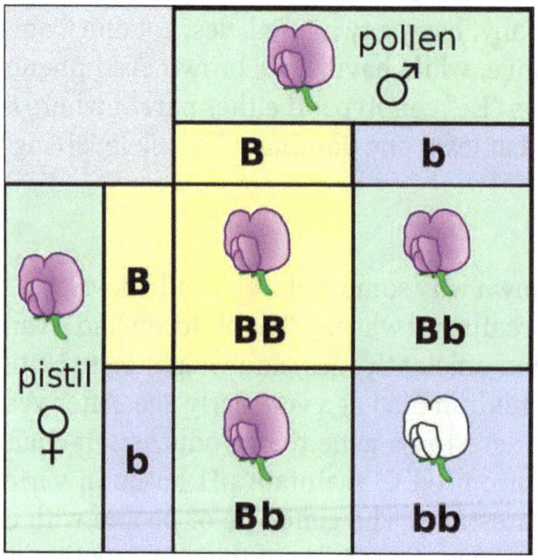

This diagram (Punnett square) shows the relation between phenotype and genotype. The petal color character of the pea plants is represented by B and b genes. B gene is the dominant trait and manifests as a purple-petalled flower. b gene is the recessive trait. A test cross between two plants that are heterozygous (Bb) for the purple petal color trait will produce 3 offspring with the purple-flower trait (BB and Bb) and 1 offspring with the white-flower trait (bb). In this example, the phenotypic ratio is 3:1.

Trait vs. Phenotype

A trait is an attribute of the organism phenotype. Thus, in genetics, the trait is also referred to as the phenotypic trait, to distinguish one trait from another within the more-inclusive term, phenotype. The phenotype of an organism is comprised of different traits. The traits may be inherited (genetically determined), acquired due to environmental influences, or result from the interaction of the two. For instance, the hair color is a character in which the traits could be black, blonde, ginger, or brunette.

Phenotype vs. Genotype

In genetics, genotype and phenotype are common terms used to explicate the appearance, function, and behavior of an organism. A genotype is defined as the set of genes that when expressed determines the characteristic or the trait of the organism. In simpler words, the genotype is the genetic contribution to the phenotype. The genes make up the sequences of DNA. In humans and other organisms, they occur in pairs – one coming from the male parent and the other from the female parent. The pairs of genes occupying the same loci on the chromosomes and controlling the same trait are called alleles. So, how are alleles and traits related? Typically, a pair of genes (or a pair of alleles) for a particular trait is comprised of two, one is dominant and the other, recessive.

The dominant allele will be expressed as a trait whereas the recessive allele will not be expressed. Annotating the dominant allele with A and the recessive allele with a, there could be three possible genotypes: (1) AA, the homozygous dominant allele, (2) Aa, the heterozygous dominant, and (3) aa, the homozygous recessive. The genotype of an organism is a major determinant of the phenotype of an organism. But how does genotype affect phenotype? Let's take for instance the pair of alleles (or pair of genes) that determines a particular trait where one of them is dominant (A) and the other, recessive (a). The dominant allele (A) will be expressed and become a part of the phenotype of the organism whereas the effect of the recessive allele (a) will be masked. When a trait follows Mendelian inheritance, the A will manifest as a trait whereas a will not. Thus, the phenotype of an organism shall include the traits of all the expressed genes. In humans though, many of the observable traits are more complex than those that follow the Mendelian pattern. Complex traits, such as height and skin color, are brought about by the interactions of multiple alleles, as in the case of polygenic inheritance.

Expression

The phenotype is the result of genetic factors, environmental influences, and random genetic variations. When the pair of alleles determining a particular trait consists of the same genes, e.g. AA or aa, the trait is described as homozygous. When the allelic composition consists of different genes, e.g. Aa, the trait is described as heterozygous. The presence of the dominant allele, i.e. AA or Aa, will result in the manifestation of the trait (A) whereas the absence of the dominant allele, i.e. aa, will result in the expression of the other trait (a). This is a case of complete dominance and it follows the Mendelian inheritance. In cases of codominance, incomplete dominance, and polygenic inheritance, the manifestation of a trait will not follow the Mendelian pattern. For instance, in codominance, the alleles of a gene pair in a heterozygote will be fully expressed since both alleles in a pair are dominant (e.g. AB). In incomplete dominance, the resulting trait will be an intermediate of the effects of the two alleles. This is because the dominant allele will only be partially expressed. As a result, the heterozygous offspring will show a phenotype that is intermediate to the phenotype of the parents.

Apart from the genetic interactions, the phenotype of an organism will also depend on the influence of the environment and random (genetic) variations. Exposure to environmental factors could affect the appearance of an organism. For instance, a light-colored skin that is constantly exposed to the sun's rays will cause the skin to darken from increased melanin production. As for random variation, the genetic mutation could alter the physical trait or at least the fitness of an organism. Changes in the genes are essential as they drive evolution and natural selection. Genotypes, environmental effects, and genetic variations help explain why individuals tend to have different phenotypes.

Extreme Phenotype

An extreme phenotype arises when the parents' alleles come together and results in a hybrid with a phenotype that is greater or higher than the phenotypes of its parents. Its transgressive phenotype may be beneficial or detrimental depending on how it affects the overall fitness of the offspring. The formation of extreme phenotypes is called transgressive segregation. An example of a hybrid with an extreme phenotype is the offspring from a cross between Helianthus annuus and Helianthus petiolaris. The two sunflower species produced hybrids that have transgressive traits. The hybrids, as opposed to their parents, are able to thrive in areas where their parents cannot. They are able to survive in sand dunes and salt marshes.

Recombinant Phenotype

Meiosis is one of the important biological events that lead to greater variation in the phenotypes of organisms. Particularly, during the metaphase of meiosis I, the homologous chromosomes come together to exchange genes via homologous recombination. As the homologous chromosomes reach the end of meiosis (telophase II), the four daughter cells will have chromosomes distinct from one another. Some of them will become gametes containing recombinant genes. When such gamete is fertilized with a wild type, for instance, it will develop into an offspring with a recombinant phenotype, i.e. a phenotype that is different from the phenotypes of its parents. How to determine the phenotype of recombinants? A test-cross on two traits (e.g. blue-bodied, normal-winged fly parent and black-bodied, vestigial-winged fly parent) can help identify recombinant phenotypes. The offspring with phenotypes different from the parents (e.g. blue-bodied, vestigial-winged fly, or black-bodied, normal-winged fly) indicate that they are recombinants.

Phenotypic Ratio

A Punnett square is another tool that can be used to identify all possible allelic combinations in a test cross. It can predict the genotypes and the phenotypes of the offspring. It is a diagram in grids and letters to represent alleles. An uppercase letter (e.g. A) denotes a dominant trait or genotype and a lowercase denotes a recessive trait or genotype (e.g. a). Based on the Punnett square, the phenotypic ratio (as well as the genotypic ratio) can be determined. A phenotypic ratio is a ratio that can be predicted from a test cross. It can be known based on the phenotypes of the offspring, i.e. the frequency of offspring manifesting different traits or trait combinations. For example, the expected phenotypic ratio of AaBb × aabb dihybrid cross (i.e. a cross that deals with two characters, e.g. body color (A for blue, a for black) and wing morphology (B for normal wings, b for vestigial wings) is 1:1:1:1, based on the four possible different phenotypes: AaBb (blue, normal-winged fly), aaBb (black, normal-winged fly), Aabb (blue, vestigial-winged fly), and aabb (black, vestigial-winged fly). The phenotype of an organism refers to the various traits an organism possesses. Examples are blue eye trait (for eye color character), brown skin trait (for skin color character), long-tail trait (for tail character), five-petalled trait (for flower character), etc. Behavior is also a phenotype example. For example, the behavior and cognitive traits of individuals with mental retardation are behavioral phenotypes.

Allele

An allele is a term coined to describe a specific copy of a gene. Genes, the DNA sequences controlling our traits, are usually found in two copies in Eukaryotic genomes; each copy (allele) is inherited from one parent. Each allele occupies a specific region on the chromosome called a locus. The gene's two alleles are located in the same region in two homologous chromosomes, one inherited from each parent.

The term "phenotype" is used to describe a visible trait or characteristic. Assuming that the trait is controlled by only one gene, both alleles constitute a genotype. Accordingly, the term genotype is used to describe an organism's set of alleles coding for each trait. Depending on each allele's DNA sequence, the genotype can be described as homozygous or heterozygous of a particular gene. Homozygous genotype means that the organism's genome has two identical alleles for a specific gene. That means both alleles contribute equally to the appearance of the trait. Heterozygous genotype, in contrast, contains two different forms of a particular gene.

One classic example explaining how alleles work is the alleles responsible for the flower color in garden peas, one of the seven traits studied by Gregor Mendel in his experiments that later established the principles of genetics. The color of garden peas flowers can be either purple or white. The gene coding for this trait, responsible for producing a purple pigment, is found in two forms, allele for *purple color* and allele for *white color.*

Alleles are designated by letters to describe genotypes. The purple color's allele (referred to as *P*) can naturally express the purple pigment while the white color allele (referred to as *p*) cannot. Hence, the resulting genotype is written as *Pp*. Other genotypes that can be found for this trait are *PP* (Purple flower) and *pp* (white flower). However, one P allele can produce enough pigment to make the flower purple color in this particular example. Therefore, plants with either *PP* or *Pp* genotypes will have purple flowers, and only plants with the pp genotype have white flowers.

In this situation, the allele that expresses the trait over the other trait is called the *dominant allele* (the P allele in this case). The other allele (p) is called *recessive* and is only expressed when present in two copies. However, this is not always the case. In some genes, two different alleles (heterozygous genotype) are equally expressed, a phenomenon called codominance. One of the most common examples of this type of inheritance is ABO blood grouping alleles.

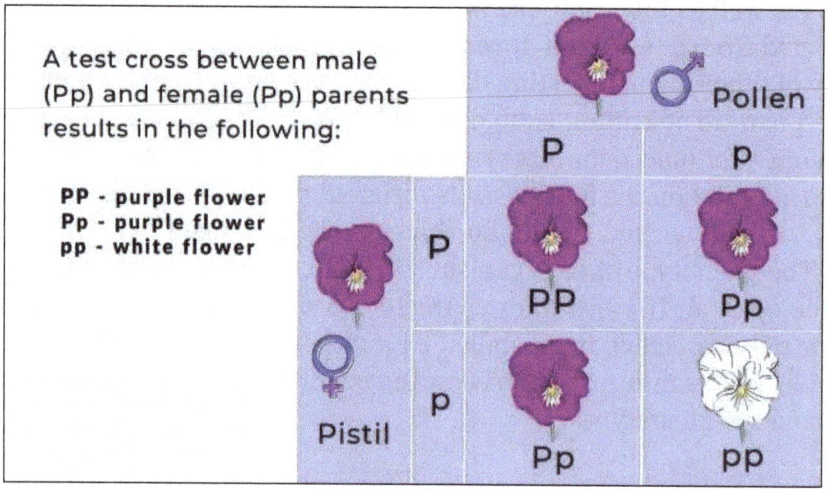

ABO locus has three allelic forms: *A, B, O*. The genotype of this gene determines the antigens on the red blood cells, hence the blood group phenotype. Both A and B alleles encode for glycosyltransferase that catalyzes the final step in the A and B antigens' synthesis pathway, while the O allele codes for an inactive version of the enzyme. The enzymes coded by A and B alleles differ in only four amino acids, due to several single nucleotide polymorphisms (SNPs) in the ABO gene. The table below shows the inheritance pattern of the ABO gene. Since the locus has three alternative alleles, six possible genotypes can result from mating. However, only four phenotypes (blood groups) occur, A, B, AB, and O. Two inheritance patterns can be seen in blood groups. Both A and B alleles are dominant when they are found with O allele. However, If a person has an AB genotype, both alleles are equally expressed (codominance), resulting in a phenotype where both A and B antigens are found on the surface of red blood cells.

Table: Inheritance pattern of the ABO gene.

ABO genotypes in the offspring A		ABO alleles inherited from the mother		
		B	O	
ABO alleles inherited from the father	A	A	AB	A
	B	AB	B	B
	O	A	B	O

Multiple Alleles

Some genes, like ABO gene, could have three or more alternative allelic forms. Many know human genes have multiple alleles. Each individual can have only two alleles of each gene. However, some genes could have hundreds of alleles within a population. Any allele results from a change in a gene's DNA sequence, even if it occurs in a single nucleotide. Therefore, multiple alleles of a given gene do not necessarily mean variation in the phenotype. Many allelic forms result from changes in the gene's sequence without affecting protein sequence or function. All these 'normal' alleles can be described as "wild-type" alleles compared to mutant alleles that affect protein structure or function.

Multiple alleles are also used to describe all mutations found in a gene in a population. For example, multiple loss-of-function mutations can be reported in a population of patients suffering from a genetic disease. Although different mutations lead to the same result, loss of protein function, each mutation constitutes a different mutant allele as long as it is located in a different gene site.

Allelic Variation

As illustrated with ABO blood groups, genetic variation between alleles could lead to a *protein variation*. Even a small variation in protein level, four amino acids, in this case, can result in drastic effects on the phenotypes. Indeed, a blood transfusion to an incompatible patient could lead to death due to eliciting the recipient's immune response to the ABO antigens present on the donor's RBCs. Therefore, it can be said that one of the significant factors contributing to the genetic variation observed between individuals is the allelic variation between their genes.

The blood group and flower color phenotypes are examples of discontinuous variation, where a trait is found in two or more distinct alternative forms. In this type of variation, the different phenotypes can be easily distinguished. Geneticists use the term *polymorphism* to describe the traits with two or more common phenotypes in a population and morphs to describe the individual phenotypes. In some cases, rare, exceptional phenotypes occur; these are called mutants, and the more common normal phenotype is called wild-type. Although both polymorphisms and mutations originate from DNA sequence changes, somehow, polymorphism changes became more common.

The second type of genetic variation is called continuous variation. In this type of variation, a trait shows a continuous range of phenotypes that cannot be distinguished as distinct phenotypes, as is the case with the discontinuous variation. Examples of characters that show continuous variation include weight, height, eye color, and similar measurable characteristics. Such traits are usually encoded by more than one gene; analyzing these phenotypes is more complicated than those showing discontinuous variation.

Molecular Basis of Allelic Variation

Phenotypes can be expressed due to a single gene's activity, as in most discontinuous phenotypes, or several genes, as in continuous phenotypes. Most genes code for proteins that directly express the trait. Being the direct products of gene expression, Proteins are the main factors presenting the phenotype by executing their cellular functions. Proteins can be receptors responsible for presenting antigens to the immune cells, pigments, hormones, antibodies, or enzymes. Genetic variation in the genes code for these proteins can have several consequences, ranging from no effect to disrupting these genes' function. Some genetic variations involve the change of a single nucleotide, the building block of DNA. Such mutation, referred to as nucleotide substitution, is the most common and simplest form of genetic variation. If a substitution occurs in a coding region within the gene, that will result in amino acid substitution.

The effect of amino acid substitution on the protein function and the phenotype depends primarily on the changed amino acid site within the protein's three-dimensional structure. For example, enzymes function is to catalyze chemical reactions by binding to their substrates through an active site, a region within the 3D structure of the enzyme that is specifically bound to a substrate in a lock-and-key manner. And since biological function depends on structure, any slight change in the active site's 3D structure will directly affect the enzymatic function. Therefore, a base-pair substitution that results in changing an amino acid in the active site or in a different region that will change the active site's 3D structure will prevent the enzyme activity. Other base substitutions can create a stop codon resulting in the premature termination of translation and the creation of truncated protein products.

Mutations can also occur in the form of an addition or deletion of one or a few nucleotides. The effect of these mutations is usually more damaging to the protein because the mRNA molecule is read in a group of three "codons"; therefore, unless the deletion or insertion involves a multiple of three nucleotides, a shift in the reading frame will result, changing all the subsequence codons. Such mutations are called "frameshift mutations."

Examples of Allele

- Beta-thalassemia and β-globin gene: Beta-thalassemia is characterized by the reduction or absence of beta-globin chains of the hemoglobin. More than 300 β-thalassemia alleles have been reported so far; most of them are inherited in a Mendelian recessive pattern. Most mutations reported involve a single or a few nucleotides within the β-globin gene or its flanking regions. Some deletion mutations have also been reported and result in the complete loss of β chain production. The severity of the disease depends mainly on the quantitative reduction level of β chains. Due to the wide range of mutations reported in the β-globin gene and the variation of phenotypes observed, Beta-thalassemias are observed as a group of inherited blood disorders.
- STR alleles and forensic identification: Short tandem repeats (STRs) are short repetitive DNA sequences (2-6 nucleotides) scattered all over the genome. They are non-coding DNA sequences but can be linked to specific genetic conditions. However, the primary and most successful application of STR markers is in forensic analysis. The number of repeats in each STR locus is highly variable between individuals. Scientists used these variations within a population and developed a test for forensic identification using multiple STR markers. The use of multiple markers in the same assay gives a high power of discrimination in the form of a unique "DNA fingerprint" that can be used for human identification in cases like crime suspects or unrecognized bodies.

Methods in Genetics

The study of genes has proved to be a powerful approach to understanding biological systems. Because genes affect virtually every aspect of the structure and function of an organism, being able to identify and determine the role of genes and the proteins that they encode is an important step in charting the various processes that underly a particular character under investigation. It is interesting that geneticists study not only hereditary mechanisms, but all biological mechanisms. Many different methodologies are used to study genes and gene activities, and these methodologies can be summarized briefly as follows:

- Isolation of mutants affecting the process under study: Each mutant gene reveals a genetic component of the process, and together they show the range of proteins that interact in that specific process.

- Analysis of progeny of controlled matings ("crosses") between mutants and other discontinuous variants: This type of analysis identifies genes and their alleles, their chromosomal locations, and their inheritance patterns.
- Biochemical analysis of cellular processes controlled by genes: Life is basically a complex set of chemical reactions; so studying the ways in which genes are relevant to these reactions is an important way of dissecting this complex chemistry. Mutant alleles underlying defective function are invaluable in this type of analysis. The basic approach is to find out how the cellular chemistry is disturbed in the mutant individual and, from this information, deduce the role of the gene. The deductions from many genes are assembled to reveal the larger picture.

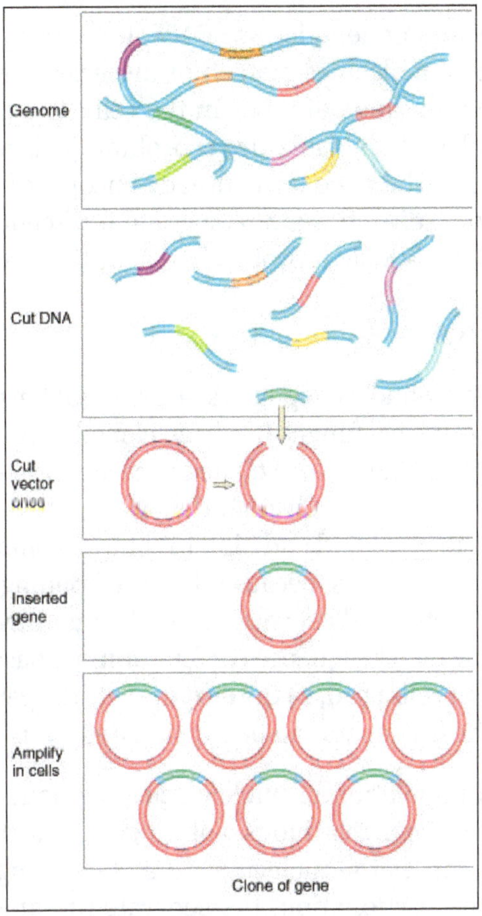

Basic cloning methodology.

- Microscopic analysis: Chromosome structure and movement have long been an integral part of genetics, but new technologies have provided ways of labeling genes and gene products so that their locations can be easily visualized under the microscope.
- Analysis of DNA directly: Because the genetic material is composed of DNA, the ultimate characterization is the analysis of DNA itself. Many procedures, including cloning, are

used. Cloning is a procedure by which an individual gene can be isolated and amplified (multiply copied) to produce a pure sample for analysis. After the clone of a gene has been obtained, its nucleotide sequence can be determined and hence important information about its structure and function can be obtained.

Cloning works by inserting the gene to be amplified into a small accessory chromosome and letting this chromosome do the job of replicating and amplifying its "passenger" fragment. This small chromosome is called a vector or carrier. Commonly used vectors are plasmids, which are nonessential extra DNA molecules found naturally in many bacteria. The DNA of any organism can be inserted into a vector; the hybrid construct is then introduced into a single bacterial cell and a large culture grows from this cell. The vector and insert can then be removed from disrupted cells and studied as appropriate.

Detecting Specific Molecules of DNA, RNA and Protein

Because the main macromolecules of genetics are DNA, RNA and protein, genetic analysis often requires the detection of specific molecules of each of these three types. How can specific molecules be identified among the thousands of types in the cell? The most extensively used method for detecting specific macromolecules in a mixture is probing. This method makes use of the specificity of intermolecular binding, which we have already encountered several times. The probe is labeled in some way, either by a radioactive atom or by a fluorescent compound, so that the site of binding can easily be detected. Let's look at probes for DNA, RNA, and protein.

Probing for a Specific DNA

A cloned gene can act as a probe for finding segments of DNA that have the same or a very similar sequence. For example, if a gene *G* from a fungus has been cloned, it might be of interest to determine whether plants have the same gene. The use of a cloned gene as a probe takes us back to the principle of base complementarity. The probe works through the principle that, in solution, the random motion of probe molecules enables them to find and bind to complementary sequences. The experiment must be done with separated DNA strands, because then the bonding sites of the bases are unoccupied. DNA from the plant is extracted and cut with one of the many available types of restriction enzymes, which cut DNA at specific target sequences of four or more bases. The target sequences are at the same positions in all the plant cells used, so the enzyme cuts the genome into defined populations of segments of specific sizes. The fragments can be fractionated by using electrophoresis.

Electrophoresis fractionates a population of nucleic acid fragments on the basis of size. The cut mixture is placed in a small well in a gelatinous slab (a gel), and the gel is placed in a powerful electrical field. The electricity causes the molecules to move through the gel at speeds inversely proportional to their size. After fractionation, the separated fragments are blotted onto a piece of porous membrane, where they maintain the same relative positions. This procedure is called a Southern blot. After having been heated to separate the DNA strands and hold the DNA in position, the membrane is placed in a solution of the probe. The single-stranded probe will find and bind to its complementary DNA sequence. For example,

TAGGCATCG	Probe
ACTAATCCATAGCTTA	Genomic fragment

On the blot, this binding concentrates the label in one spot, as shown in figure.

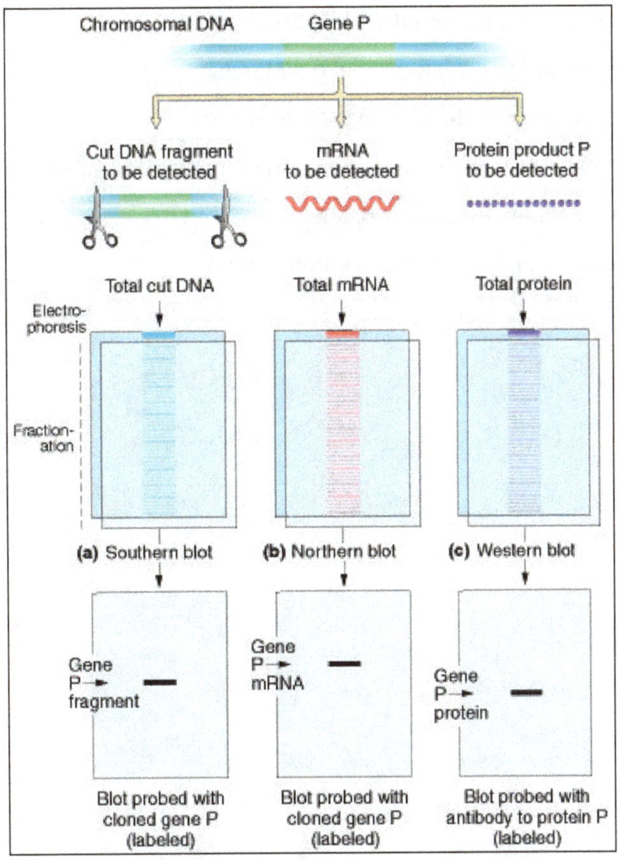

Probing DNA, RNA, and protein mixtures.

Probing for a Specific RNA

It is often necessary to determine whether a gene is being transcribed in some particular tissue. This determination can be accomplished by using a modification of the Southern analysis. Total mRNA is extracted from the tissue, fractionated electrophoretically, and blotted onto a membrane (this procedure is called a Northern blot). The cloned gene is used as a probe and its label will highlight the mRNA in question if it is present.

Probing for a Specific Protein

Probing for proteins is generally performed with antibodies because an antibody has a specific lock-and-key fit with its antigen. The protein mixture is electrophoresed and blotted onto a membrane (this procedure is a Western blot). The position of a specific protein on the membrane is revealed by bathing the membrane in a solution of antibody obtained from a rabbit into which the protein has been injected. The position of the protein is revealed by the position of the label that the antibody carries.

References

- Genetics: biologydictionary.net, Retrieved 21, July 2020
- Nucleic-acid, science: britannica.com, Retrieved 09, March 2020
- Nucleotide: biologydictionary.net, Retrieved 13, January 2020
- Important-notes-of-biology-for-neet-chromosome-structure: byjus.com, Retrieved 18, August 2020
- Heredity: biologydictionary.net, Retrieved 26, May 2020
- Phenotype: biologyonline.com, Retrieved 18, February 2020

2

Principles of Genetics

Principles of genetics include law of inheritance, natural selection, chromosome abnormality, genetic variation, etc. Natural selection is a key mechanism of evolution that describes the differences in the survival and reproduction of individuals having different phenotypes. All these diverse principles of genetics have been carefully analyzed in this chapter.

Law of Inheritance

Mendelian inheritance (or Mendelian genetics or Mendelism) is a set of primary tenets relating to the transmission of hereditary characteristics from parent organisms to their children; it underlies much of genetics. The tenets were initially derived from the work of Gregor Mendel published in 1865 and 1866, which was "re-discovered" in 1900; they were initially very controversial, but they soon became the core of classical genetics. The laws of inheritance were derived by Gregor Mendel, a 19th century monk conducting hybridization experiments in garden peas (*Pisum sativum*). Between 1856 and 1863, he cultivated and tested some 28,000 pea plants. From these experiments, he deduced two generalizations that later became known as Mendel's Laws of Heredity or Mendelian inheritance.

Mendel's Laws

Mendel discovered that by crossing true-breeding white flower and true-breeding purple flower plants, the result was a hybrid offspring. Rather than being a mix of the two colors, the offspring was purple flowered. He then conceived the idea of heredity units, which he called "factors", one of which is a recessive characteristic and the other dominant. Mendel said that factors, later called genes, normally occur in pairs in ordinary body cells, yet segregate during the formation of sex cells. Each member of the pair becomes part of the separate sex cell. The dominant gene, such as the purple flower in Mendel's plants, will hide the recessive gene, the white flower. After Mendel self-fertilized the F1 generation and obtained an F2 generation with a 3:1 ratio, he correctly theorized that genes can be paired in three different ways for each trait: AA, aa, and Aa. The capital A represents the dominant factor while the lowercase a represents the recessive.

Mendel stated that each individual has two alleles for each trait, one from each parent. Thus, he formed the "first rule", the Law of Segregation, which states individuals, possess two alleles and a parent passes only one allele to his/her offspring. One allele is given by the female parent and the other is given by the male parent. The two factors may or may not contain the same information. If the two alleles are identical, the individual is called homozygous for the trait. If the two alleles are different, the individual is called heterozygous. The presence of an allele does not promise that the trait will be expressed in the individual that possesses it. In heterozygous individuals, the only allele that is expressed is the dominant. The recessive allele is present, but its expression is hidden. The genotype of an individual is made up of the many alleles it possesses. An individual's physical appearance, or phenotype, is determined by its alleles as well as by its environment.

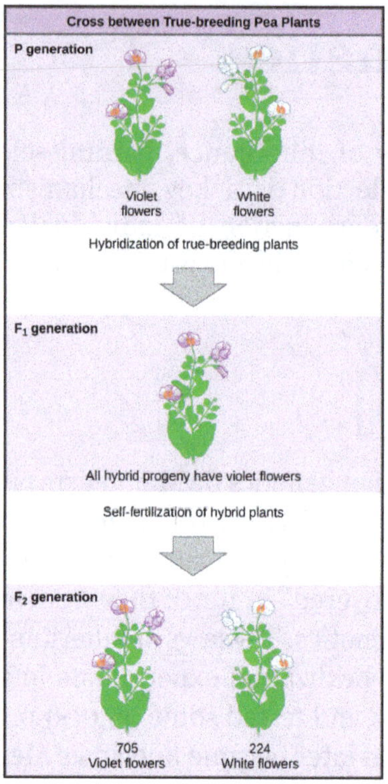

Mendel's Pea Plants: In one of his experiments on inheritance patterns, Mendel crossed plants that were true-breeding for violet flower color with plants true-breeding for white flower color (the P generation). The resulting hybrids in the F1 generation all had violet flowers. In the F2 generation, approximately three-quarters of the plants had violet flowers, and one-quarter had white flowers.

Mendel also analyzed the pattern of inheritance of seven pairs of contrasting traits in the domestic pea plant. He did this by cross-breeding dihybrids; that is, plants that were heterozygous for the alleles controlling two different traits. Mendel then crossed these dihybrids. If it is inevitable that round seeds must always be yellow and wrinkled seeds must be green, then he would have expected that this would produce a typical monohybrid cross: 75 percent round-yellow; 25 percent wrinkled-green. But, in fact, his mating generated seeds that showed all possible combinations of the color and texture traits. He found 9/16 of the offspring were round-yellow, 3/16 were round-green, 3/16 were wrinkled-yellow, and 1/16 were wrinkled-green. Finding in every case that each of his seven traits was inherited independently of the others, he formed his "second rule", the Law of Independent Assortment, which states the inheritance of one pair of factors (genes) is independent of the inheritance of the other pair. Today we know that this rule holds only if the genes are on separate chromosomes.

Mendel's Law of Dominance

In a heterozygote, the allele which masks the other is referred to as dominant, while the allele that is masked is referred to as recessive.

Alleles can be Dominant or Recessive

Most familiar animals and some plants have paired chromosomes and are described as diploid. They have two versions of each chromosome: one contributed by the female parent in her ovum and one by the male parent in his sperm. These are joined at fertilization. The ovum and sperm cells (the gametes) have only one copy of each chromosome and are described as haploid.

Recessive traits are only visible if an individual inherits two copies of the recessive allele: The child in the photo expresses albinism, a recessive trait.

Mendel's law of dominance states that in a heterozygote, one trait will conceal the presence of another trait for the same characteristic. Rather than both alleles contributing to a phenotype, the dominant allele will be expressed exclusively. The recessive allele will remain "latent," but will be transmitted to offspring by the same manner in which the dominant allele is transmitted. The recessive trait will only be expressed by offspring that have two copies of this allele; these offspring will breed true when self-crossed.

By definition, the terms dominant and recessive refer to the genotypic interaction of alleles in producing the phenotype of the heterozygote. The key concept is genetic: which of the two alleles present in the heterozygote is expressed, such that the organism is phenotypically identical to one of the two homozygotes. It is sometimes convenient to talk about the trait corresponding to the dominant allele as the dominant trait and the trait corresponding to the hidden allele as the recessive trait.

However, this can easily lead to confusion in understanding the concept as phenotypic. For example, to say that "green peas" dominate "yellow peas" confuses inherited genotypes and expressed phenotypes. Dominance is not inherent. One allele can be dominant to a second allele, recessive to a third allele, and codominant to a fourth. If a genetic trait is recessive, a person needs to inherit two copies of the gene for the trait to be expressed. Thus, both parents have to be carriers of a recessive trait in order for a child to express that trait. Since Mendel's experiments with pea plants, other researchers have found that the law of dominance does not always hold true. Instead, several different patterns of inheritance have been found to exist.

Mendel's Law of Segregation

Mendel's Law of Segregation states that a diploid organism passes a randomly selected allele for a trait to its offspring, such that the offspring receives one allele from each parent.

Equal Segregation of Alleles

Observing that true-breeding pea plants with contrasting traits gave rise to F1 generations that all expressed the dominant trait and F2 generations that expressed the dominant and recessive traits in a 3:1 ratio, Mendel proposed the law of segregation. The law of segregation states that each individual that is a diploid has a pair of alleles (copy) for a particular trait. Each parent passes an allele at random to their offspring resulting in a diploid organism. The allele that contains the dominant trait determines the phenotype of the offspring. In essence, the law states that copies of genes separate or segregate so that each gamete receives only one allele.

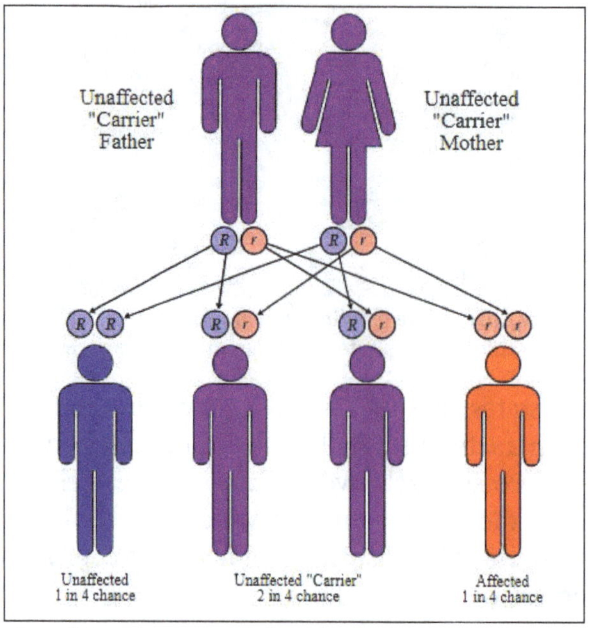

The Law of Segregation states that alleles segregate randomly into gametes: When gametes are formed, each allele of one parent segregates randomly into the gametes, such that half of the parent's gametes carry each allele.

For the F2 generation of a monohybrid cross, the following three possible combinations of genotypes could result: homozygous dominant, heterozygous, or homozygous recessive. Because heterozygotes could arise from two different pathways (receiving one dominant and one recessive allele from either parent), and because heterozygotes and homozygous dominant individuals are phenotypically identical, the law supports Mendel's observed 3:1 phenotypic ratio. The equal segregation of alleles is the reason we can apply the Punnett square to accurately predict the offspring of parents with known genotypes.

The physical basis of Mendel's law of segregation is the first division of meiosis in which the homologous chromosomes with their different versions of each gene are segregated into daughter nuclei. The behavior of homologous chromosomes during meiosis can account for the segregation of the alleles at each genetic locus to different gametes. As chromosomes separate into different gametes during meiosis, the two different alleles for a particular gene also segregate so that each gamete acquires one of the two alleles. In Mendel's experiments, the segregation and the independent assortment during meiosis in the F1 generation give rise to the F2 phenotypic ratios observed by Mendel. The role of the meiotic segregation of chromosomes in sexual reproduction was not understood by the scientific community during Mendel's lifetime.

Mendel's Law of Independent Assortment

Independent assortment allows the calculation of genotypic and phenotypic ratios based on the probability of individual gene combinations.

Independent Assortment

Mendel's law of independent assortment states that genes do not influence each other with regard to the sorting of alleles into gametes: every possible combination of alleles for every gene is equally likely to occur. The independent assortment of genes can be illustrated by the dihybrid cross: a cross between two true-breeding parents that express different traits for two characteristics. Consider the characteristics of seed color and seed texture for two pea plants: one that has green, wrinkled seeds (yyrr) and another that has yellow, round seeds (YYRR). Because each parent is homozygous, the law of segregation indicates that the gametes for the green/wrinkled plant all are yr, while the gametes for the yellow/round plant are all YR. Therefore, the F1 generation of offspring all are YyRr.

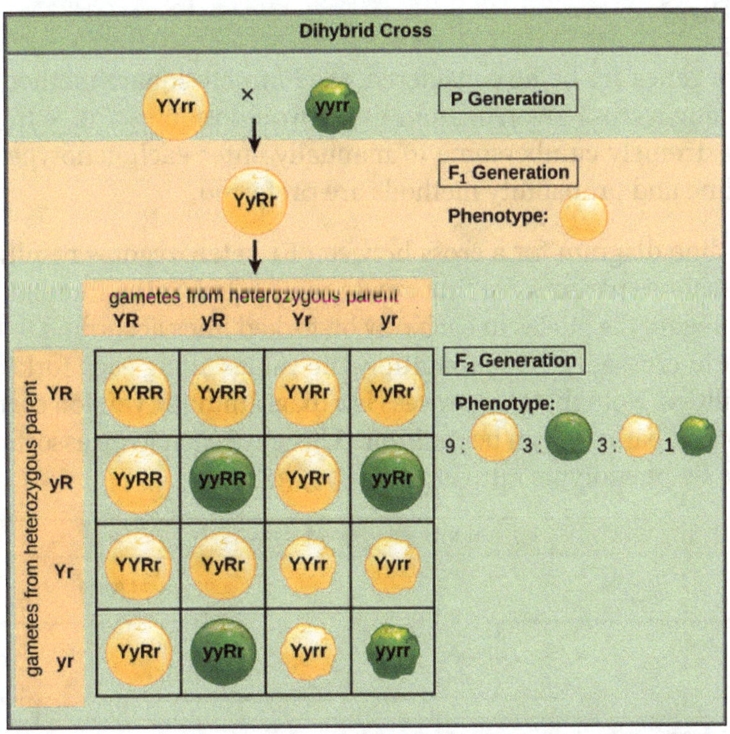

Independent assortment of 2 genes: This dihybrid cross of pea plants involves the genes for seed color and texture.

For the F2 generation, the law of segregation requires that each gamete receive either an R allele or an r allele along with either a Y allele or a y allele. The law of independent assortment states that a gamete into which an r allele sorted would be equally likely to contain either a Y allele or a y allele. Thus, there are four equally likely gametes that can be formed when the YyRr heterozygote is self-crossed as follows: YR, Yr, yR, and yr. Arranging these gametes along the top and left of a 4 × 4 Punnett square gives us 16 equally likely genotypic combinations. From these genotypes, we infer a phenotypic ratio of 9 round/yellow:3 round/green:3 wrinkled/yellow:1 wrinkled/green. These are the offspring ratios we would expect, assuming we performed the crosses with a large enough sample size.

Because of independent assortment and dominance, the 9:3:3:1 dihybrid phenotypic ratio can be collapsed into two 3:1 ratios, characteristic of any monohybrid cross that follows a dominant and recessive pattern. Ignoring seed color and considering only seed texture in the above dihybrid cross, we would expect that three-quarters of the F_2 generation offspring would be round and one-quarter would be wrinkled. Similarly, isolating only seed color, we would assume that three-quarters of the F_2 offspring would be yellow and one-quarter would be green. The sorting of alleles for texture and color are independent events, so we can apply the product rule. Therefore, the proportion of round and yellow F_2 offspring is expected to be (3/4) × (3/4) = 9/16, and the proportion of wrinkled and green offspring is expected to be (1/4) × (1/4) = 1/16. These proportions are identical to those obtained using a Punnett square. Round/green and wrinkled/yellow offspring can also be calculated using the product rule as each of these genotypes includes one dominant and one recessive phenotype. Therefore, the proportion of each is calculated as (3/4) × (1/4) = 3/16.

Forked-line Method

When more than two genes are being considered, the Punnett-square method becomes unwieldy. For instance, examining a cross involving four genes would require a 16 × 16 grid containing 256 boxes. It would be extremely cumbersome to manually enter each genotype. For more complex crosses, the forked-line and probability methods are preferred.

To prepare a forked-line diagram for a cross between F1 heterozygotes resulting from a cross between AABBCC and aabbcc parents, we first create rows equal to the number of genes being considered and then segregate the alleles in each row on forked lines according to the probabilities for individual monohybrid crosses. We then multiply the values along each forked path to obtain the F2 offspring probabilities. Note that this process is a diagrammatic version of the product rule. The values along each forked pathway can be multiplied because each gene assorts independently. For a trihybrid cross, the F2 phenotypic ratio is 27:9:9:9:3:3:3:1.

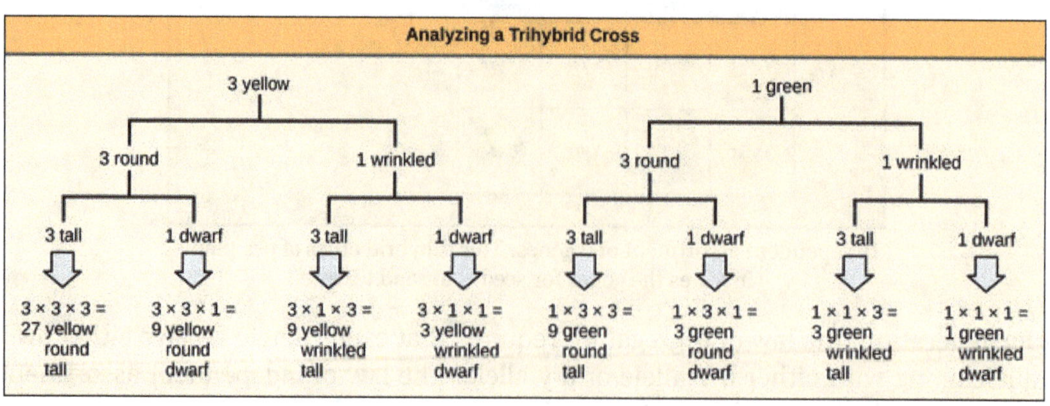

Independent assortment of 3 genes: The forked-line method can be used to analyze a trihybrid cross. Here, the probability for color in the F_2 generation occupies the top row (3 yellow:1 green). The probability for shape occupies the second row (3 round: 1 wrinked), and the probability for height occupies the third row (3 tall:1 dwarf). The probability for each possible combination of traits is calculated by multiplying the probability for each individual trait. Thus, the probability of F2 offspring having yellow, round, and tall traits is 3 × 3 × 3, or 27.

Probability Method

While the forked-line method is a diagrammatic approach to keeping track of probabilities in a cross, the probability method gives the proportions of offspring expected to exhibit each phenotype (or genotype) without the added visual assistance.

To fully demonstrate the power of the probability method, however, we can consider specific genetic calculations. For instance, for a tetrahybrid cross between individuals that are heterozygotes for all four genes, and in which all four genes are sorting independently in a dominant and recessive pattern, what proportion of the offspring will be expected to be homozygous recessive for all four alleles? Rather than writing out every possible genotype, we can use the probability method. We know that for each gene the fraction of homozygous recessive offspring will be 1/4. Therefore, multiplying this fraction for each of the four genes, (1/4) × (1/4) × (1/4) × (1/4), we determine that 1/256 of the offspring will be quadruply homozygous recessive.

Genetic Linkage and Violation of the Law of Independent Assortment

TtIi × TtIi	TI	Ti	tI	ti
TI	TTII	TTIi	TtII	TtIi
Ti	TTIi	TTii	TtIi	Ttii
tI	TtII	TtIi	ttII	ttIi
ti	TtIi	Ttii	ttIi	ttii

Unlinked genes assort independently: This figure shows all possible combinations of offspring resulting from a dihybrid cross of pea plants that are heterozygous for the tall/dwarf and inflated/constricted alleles.

Genes that are on the same chromosome, or "linked", do not assort independently, but can be separated by recombination. Although all of Mendel's pea characteristics behaved according to the law of independent assortment, we now know that some allele combinations are not inherited independently of each other. Genes that are located on separate non-homologous chromosomes will always sort independently. However, each chromosome contains hundreds or thousands of genes organized linearly on chromosomes like beads on a string. The segregation of alleles into gametes can be influenced by linkage, in which genes that are located physically close to each other on the same chromosome are more likely to be inherited as a pair. However, because of the process of recombination, or "crossover," it is possible for two genes on the same chromosome to behave independently, or as if they are not linked. To understand this, let's consider the biological basis of gene linkage and recombination.

Homologous chromosomes possess the same genes in the same linear order. The alleles may differ on homologous chromosome pairs, but the genes to which they correspond do not. In preparation for the first division of meiosis, homologous chromosomes replicate and synapse. Like genes on the homologs align with each other. At this stage, segments of homologous chromosomes exchange linear segments of genetic material. This process is called recombination, or crossover, and it is a common genetic process. Because the genes are aligned during recombination, the gene order is not altered. Instead, the result of recombination is that maternal and paternal alleles are combined onto the same chromosome. Across a given chromosome, several recombination events may occur, causing extensive shuffling of alleles.

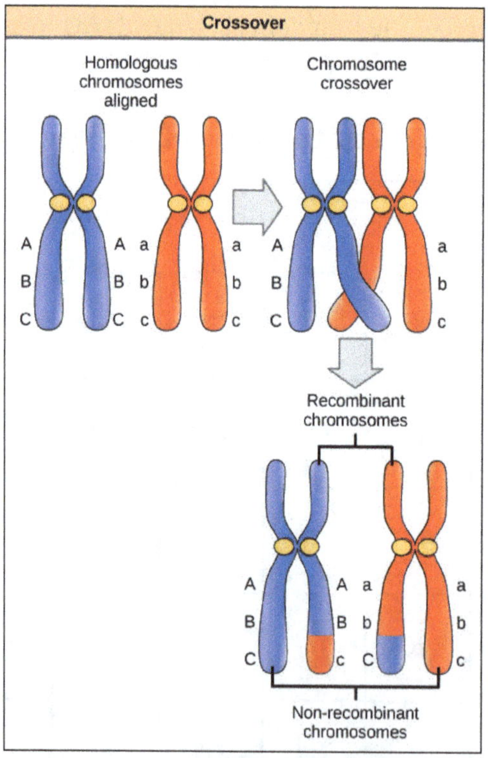

Linked genes can be separated by recombination: The process of crossover, or recombination, occurs when two homologous chromosomes align during meiosis and exchange a segment of genetic material. Here, the alleles for gene C were exchanged. The result is two recombinant and two non-recombinant chromosomes.

When two genes are located in close proximity on the same chromosome, they are considered linked, and their alleles tend to be transmitted through meiosis together. To exemplify this, imagine a dihybrid cross involving flower color and plant height in which the genes are next to each other on the chromosome. If one homologous chromosome has alleles for tall plants and red flowers, and the other chromosome has genes for short plants and yellow flowers, then when the gametes are formed, the tall and red alleles will go together into a gamete and the short and yellow alleles will go into other gametes. These are called the parental genotypes because they have been inherited intact from the parents of the individual producing gametes. But unlike if the genes were on different chromosomes, there will be no gametes with tall and yellow alleles and no gametes with short and red alleles. If you create the Punnett square with these gametes, you will see that the classical Mendelian prediction of a 9:3:3:1 outcome of a dihybrid cross would not apply. As the distance between two genes increases, the probability of one or more crossovers between them increases, and the genes behave more like they are on separate chromosomes. Geneticists have used the proportion of recombinant gametes (the ones not like the parents) as a measure of how far apart genes are on a chromosome. Using this information, they have constructed elaborate maps of genes on chromosomes for well-studied organisms, including humans.

Mendel's seminal publication makes no mention of linkage, and many researchers have questioned whether he encountered linkage, but chose not to publish those crosses out of concern that they would invalidate his independent assortment postulate. The garden pea has seven chromosomes and some have suggested that his choice of seven characteristics was not a coincidence. However, even if the genes he examined were not located on separate chromosomes, it is possible that he simply did not observe linkage because of the extensive shuffling effects of recombination.

Epistasis

Epistasis occurs when one gene masks or interferes with the expression of another. Mendel's studies in pea plants implied that the sum of an individual's phenotype was controlled by genes (or as he called them, unit factors): every characteristic was distinctly and completely controlled by a single gene. In fact, single observable characteristics are almost always under the influence of multiple genes (each with two or more alleles) acting in unison. For example, at least eight genes contribute to eye color in humans.

In some cases, several genes can contribute to aspects of a common phenotype without their gene products ever directly interacting. In the case of organ development, for instance, genes may be expressed sequentially, with each gene adding to the complexity and specificity of the organ. Genes may function in complementary or synergistic fashions: two or more genes need to be expressed simultaneously to affect a phenotype. Genes may also oppose each other with one gene modifying the expression of another.

In epistasis, the interaction between genes is antagonistic: one gene masks or interferes with the expression of another. "Epistasis" is a word composed of Greek roots that mean "standing upon." The alleles that are being masked or silenced are said to be hypostatic to the epistatic alleles that are doing the masking. Often the biochemical basis of epistasis is a gene pathway in which the expression of one gene is dependent on the function of a gene that precedes or follows it in the pathway.

An example of epistasis is pigmentation in mice. The wild-type coat color, agouti (AA), is dominant to solid-colored fur (aa). However, a separate gene (C) is necessary for pigment production. A mouse with a recessive c allele at this locus is unable to produce pigment and is albino regardless of the allele present at locus A. Therefore, the genotypes AAcc, Aacc, and aacc all produce the same albino phenotype. A cross between heterozygotes for both genes (AaCc × AaCc) would generate offspring with a phenotypic ratio of 9 agouti:3 solid color:4 albino. In this case, the C gene is epistatic to the A gene.

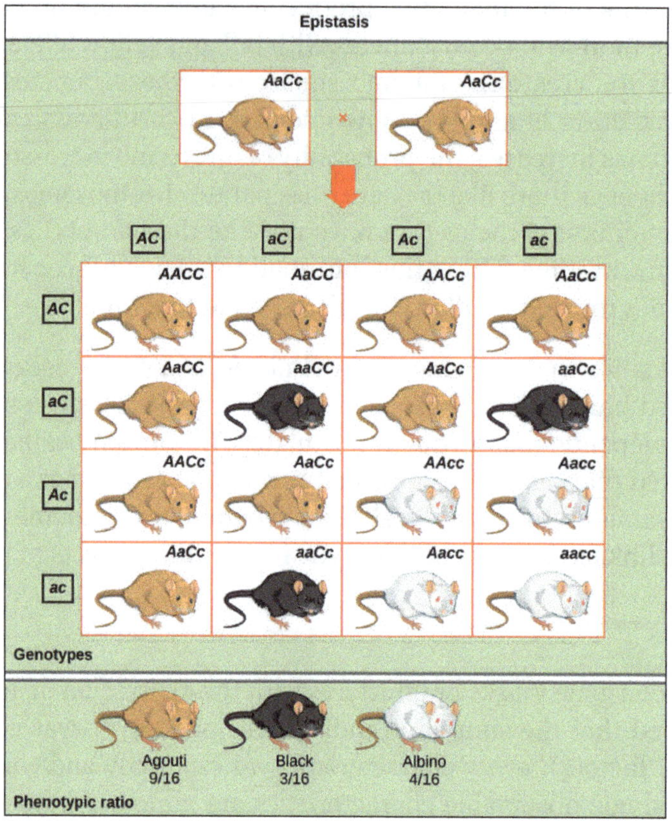

Epistasis in mouse coat color: In mice, the mottled agouti coat color (A) is dominant to a solid coloration, such as black or gray. A gene at a separate locus (C) is responsible for pigment production. The recessive c allele does not produce pigmentnand a mouse with the homozygous recessive cc genotype is albino regardless of the allele present at the A locus. Thus, the C gene is epistatic to the A gene.

Epistasis can also occur when a dominant allele masks expression at a separate gene. Fruit color in summer squash is expressed in this way. Homozygous recessive expression of the W gene (ww) coupled with homozygous dominant or heterozygous expression of the Y gene (YY or Yy) generates yellow fruit, while the wwyy genotype produces green fruit. However, if a dominant copy of the W gene is present in the homozygous or heterozygous form, the summer squash will produce white fruit regardless of the Y alleles. A cross between white heterozygotes for both genes (WwYy × WwYy) would produce offspring with a phenotypic ratio of 12 white: 3 yellow: 1 green.

Finally, epistasis can be reciprocal: either gene, when present in the dominant (or recessive) form, expresses the same phenotype. In the shepherd's purse plant (*Capsella bursa-pastoris*), the characteristic of seed shape is controlled by two genes in a dominant epistatic relationship. When the genes A and B are both homozygous recessive (aabb), the seeds are ovoid. If the dominant allele for either of these genes is present, the result is triangular seeds. That is, every possible genotype other than aabb results in triangular seeds; a cross between heterozygotes for both genes (AaBb x AaBb) would yield offspring with a phenotypic ratio of 15 triangular: 1 ovoid.

Keep in mind that any single characteristic that results in a phenotypic ratio that totals 16 is typical of a two-gene interaction. Recall the phenotypic inheritance pattern for Mendel's dihybrid cross, which considered two non-interacting genes: 9:3:3:1. Similarly, we would expect interacting gene pairs to also exhibit ratios expressed as 16 parts. Note that we are assuming the interacting genes are not linked; they are still assorting independently into gametes.

Deletion

The process of spontaneously occurring deletion must include two chromosome breaks to cut out the intervening segment. If the two ends join and one of them bears the centromere, a shortened chromosome results, which is said to carry a deletion. The deleted fragment is acentric; consequently it is immobile and will be lost. An effective mutagen for inducing chromosomal rearrangements of all kinds is ionizing radiation. This kind of radiation, of which X rays and γ rays are examples, is highly energetic and causes chromosome breaks. The way in which the breaks rejoin determines the kind of rearrangement produced. Two types of deletion are possible. Two breaks can produce an interstitial deletion, as shown in figure. In principle, a single break can cause a terminal deletion; but, because of the need for the special chromosome tips (telomeres), it is likely that apparently terminal deletions include two breaks, one close to the telomere.

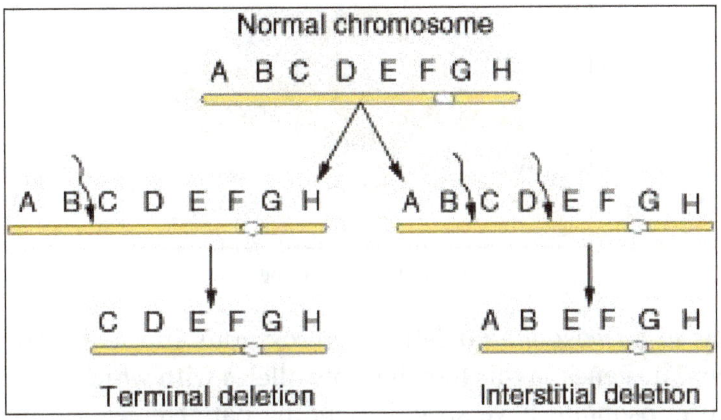

Terminal and interstitial deletions. Chromosome can be broken when struck by ionizing radiation (wavy arrows). A terminal deletion is the loss of the end of a chromosome. An interstitial deletion results after two breaks are induced if the terminal part (AB) rejoins the main body of the chromosome, with the acentric fragment (CD) being lost.

The effects of deletions depend on their size. A small deletion within a gene, called an intragenic deletion, inactivates the gene and has the same effect as other null mutations of that gene. If the homozygous null phenotype is viable (as, for example, in human albinism), then the homozygous deletion also will be viable. Intragenic deletions can be distinguished from single nucleotide changes because they are non-revertible.

Here we shall be dealing with multigenic deletions, those that remove from two to several thousand genes. Multigenic deletions have severe consequences. If by inbreeding such a deletion is made homozygous (that is, if both homologs have the same deletion), then the combination is almost always lethal. This outcome suggests that most regions of the chromosomes are essential for normal viability and that complete elimination of any segment from the genome is deleterious. Even individuals heterozygous for a multigenic deletion—those with one normal homolog and one that carries the deletion—may not survive. There are several possible reasons for this failure to survive. First, a genome has been "fine-tuned" during evolution to require a specific balance of genes, and the deletion upsets this balance. Several different types of chromosome mutations upset the ratio, or balance, of genes in a genome. Second, in many organisms there are recessive lethal and other deleterious mutations throughout the genome. If "covered" by wild-type alleles on the other homolog, these recessives are not expressed. However, a deletion can "uncover" recessives, allowing their expression at the phenotypic level.

Nevertheless, some small deletions are viable in combination with a normal homolog. In these cases, the deletion can sometimes be identified by cytogenetic analysis. If meiotic chromosomes are examined in an individual carrying a heterozygous deletion, the region of the deletion can be determined by the failure of the corresponding segment on the normal homolog to pair, resulting in a deletion loop. In insects, deletion loops are detected in the polytene chromosomes, in which the homologs are fused. A deletion can be assigned to a specific chromosome location by determining which chromosome shows the deletion loop and the position of the loop along the chromosome.

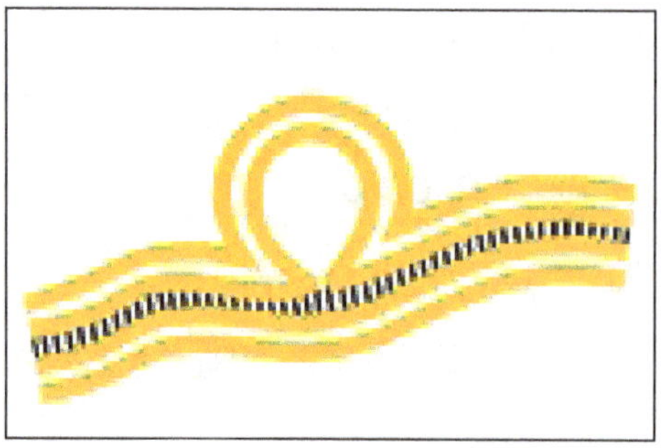

Meiotic chromosomes.

Looped configurations in a *Drosophila* deletion heterozygote. In the meiotic pairing, the normal homolog forms a loop. The genes in this loop have no alleles with which to synapse. Because polytene chromosomes in *Drosophila* have specific banding patterns, we can infer which bands are missing from the homolog with the deletion by observing which bands appear in the loop of the normal homolog.

Deletions of some chromosomal regions produce their own unique phenotypes. A good example is a deletion of one specific small chromosome region of *Drosophila*. When one homolog carries the deletion, the fly shows a unique notch-wing phenotype, so the deletion acts as a dominant mutation in this regard. But the deletion is lethal when homozygous and therefore acts as a recessive in regard to its lethal effect. The specific dominant phenotypic effect of certain deletions might be caused by one of the chromosome breaks being inside a gene, which, when disrupted, will act as a dominant mutation. What are the genetic properties of deletions? In addition to cytogenetic criteria, there are several purely genetic criteria for inferring the presence of a deletion. These criteria are particularly useful in species whose chromosomes are not easily analyzed cytogenetically.

Two genetic criteria we have encountered already. The first is the failure of the chromosome to survive as a homozygote; however, this effect could also be produced by any lethal mutation. Second, chromosomes with deletions can never revert to a normal condition. This criterion is useful only if there is some specific phenotype associated with the deletion. A third criterion is that, in heterozygous deletions, recombinant frequencies between genes flanking the deletion are lower than in control crosses. This makes intuitive sense because part of the region contains an unpaired chromosomal region, which cannot participate in crossing-over. We will see that inversions have a similar effect on recombinant frequencies but can be distinguished in other ways.

A fourth criterion for inferring the presence of a deletion is that deletion of a segment on one homolog sometimes unmasks recessive alleles present on the other homolog, leading to their unexpected expression. Consider, for example, the deletion shown in the following diagram:

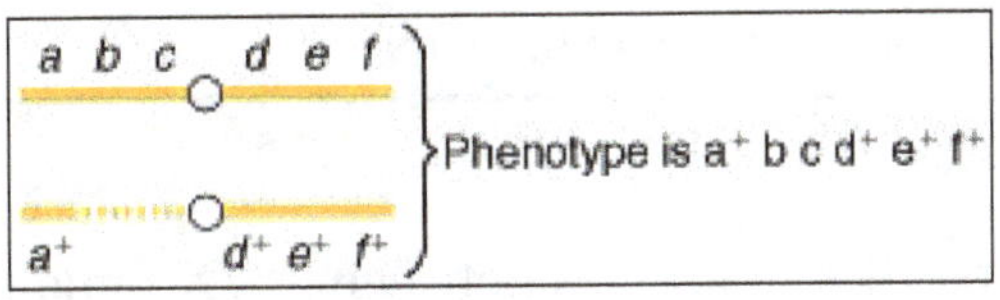

In this case, none of the six recessive alleles is expected to be expressed, but, if *b* and *c* are expressed, then a deletion is suggested to have occurred on the other homolog spanning the b^+ and c^+ loci. Because in such cases it seems that recessive alleles are showing dominance, the effect is called pseudodominance.

The pseudodominance effect can also be used in the opposite direction. A known set of overlapping deletions is used to locate the map positions of new mutant alleles. This procedure is called deletion mapping. An example from the fruit fly *Drosophila* is shown in figure. In this diagram, the recombination map is shown at the top, marked with distances in map units from the left end. The horizontal bars below the chromosome show the extent of the deletions identified at the left. The mutation prune (*pn*), for example, shows pseudodominance only with deletion 264-38, which determines its location in the 2D-4 to 3A-2 region. However, *fa* shows pseudodominance with all but two deletions, so its position can be pinpointed to band 3C-7.

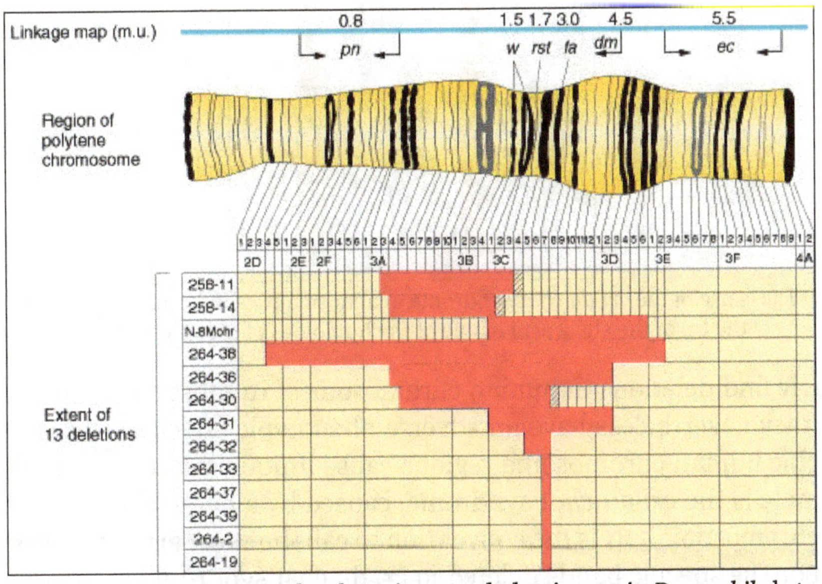

Locating genes to chromosomal regions by observing pseudodominance in *Drosophila* heterozygous for deletion and normal chromosomes. The red bars show the extent of the deleted segments in 13 deletions. All recessive alleles spanned by a deletion will be expressed.

Deletion analysis makes it possible to compare a linkage map based on recombinant frequency with the chromosome map based on deletion mapping. By and large, where this comparison has been made, the maps correspond well—a satisfying cytological endorsement of a purely genetic creation.

Moreover, pseudodominance can be used to map a small deletion that cannot be visualized microscopically. Let's consider an X chromosome in *Drosophila* that carries a recessive lethal suspected of being a deletion; we call this chromosome "X*." We can cross X*-bearing females with males carrying recessive alleles of loci on that chromosome. For example, a map of loci in the tip region is as shown below:

Suppose we obtain all wild-type flies in crosses between X*/X females and males carrying *y, dor, br, gt, rst,* and *vt* but obtain pseudodominance of *swa* and *w* with X* (that is, X*/*swa* shows the recessive swa phenotype and X*/*w* shows the recessive w phenotype). Then we have good genetic evidence for a deletion of the chromosome that includes at least the *swa* and *w* loci but not *gt* or *rst*.

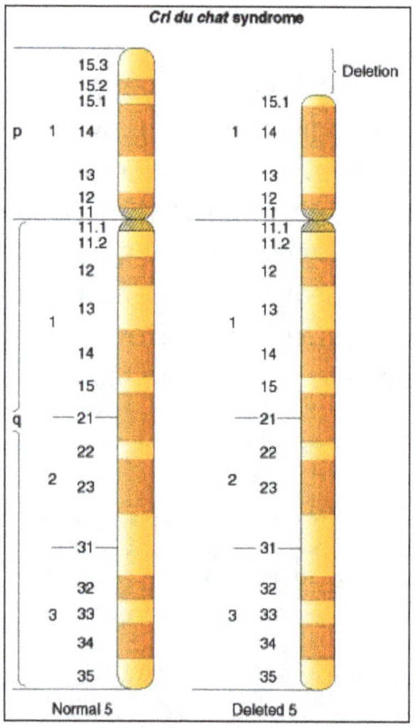

The cause of the *cri du chat* syndrome of abnormalities in humans is loss of the tip of the short arm of one of the homologs of chromosome 5.

Clinicians regularly find deletions in human chromosomes. In most cases, the deletions are relatively small, but they nevertheless have an adverse phenotypic effect, even though heterozygous. Deletions of specific human chromosome regions cause unique syndromes of phenotypic abnormalities. An example is the *cri du chat* syndrome, caused by a heterozygous deletion of the tip of the short arm of chromosome 5. It is the convention to call the short arm of a chromosome p and to call the long arm q. The specific bands deleted in cri du chat syndrome are 5p15.2 and 5p15.3, the two most distal bands identifiable on 5p. The most characteristic phenotype in the syndrome is the one that gives it its name, the distinctive catlike mewing cries made by infants with this deletion.

Other phenotypic manifestations of the syndrome are microencephaly (abnormally small head) and a moonlike face. Like syndromes caused by other deletions, the cri du chat syndrome also includes mental retardation.

Most human deletions, such as those that we have just considered, arise spontaneously in the germ line of a normal parent of an affected person; thus no signs of the deletions are found in the somatic chromosomes of the parents. However, some human deletions are produced by meiotic irregularities in a parent heterozygous for another type of rearrangement. Cri du chat syndrome, for example, can result from a parent heterozygous for a translocation. Geneticists have mapped human genes from deletions by using a molecular technique called in situ hybridization. If an interesting gene or other DNA fragment has been isolated with the use of modern molecular technology, it can be tagged with a radioactive or chemical label and then added to a chromosomal preparation under the microscope.

In such a situation, the DNA recognizes and physically binds to its normal chromosomal counterpart by nucleotide pairing and is recognized as a spot of radioactivity or dye. The precise location of such spots is difficult to correlate with specific bands, but the deletion technique comes to the rescue. If a deletion happens to span the locus in question, no spot will appear when the test is run with the chromosome carrying the deletion, because the region for binding simply is not present. By saving cell lines from patients with deletions, geneticists develop test panels of overlapping deletions spanning specific chromosomal regions, and these test panels can be used to pinpoint a gene's position. An example from chromosome 11 is shown in figure. The extent of the deletions in the test panel are shown as vertical bars, and the coded DNA fragments under test are shown at the right. If fragment 270, for example, failed to bind to deletions 35, 8, 10, 7, 9, 23, 24, A2, 27A, and 4D but did bind to the other deletions, it can be inferred that this piece of DNA originally came from the region spanned by 11q13.5 and 11q21.

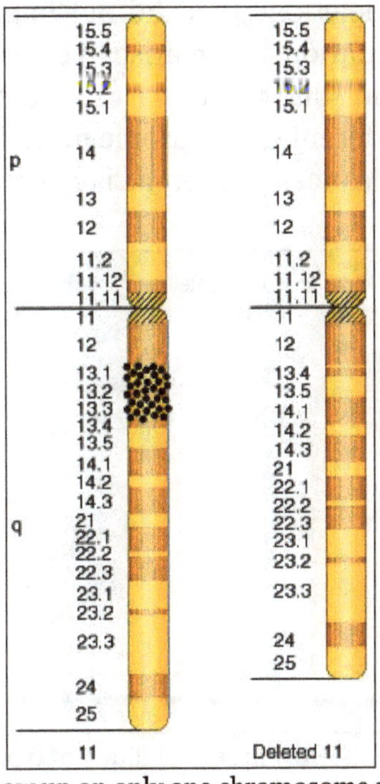

Radioactive spots show up on only one chromosome 11, because the other one has a deletion in the region where the radioactive DNA binds.

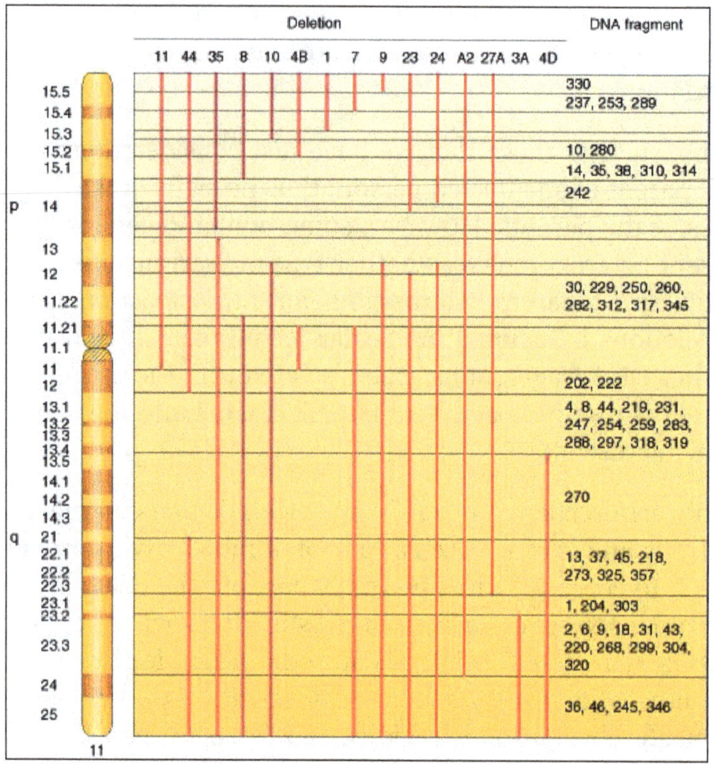

Human DNA fragments mapped to regions of chromosome 11 by their failure to bind to particular deletions. The red bars show the extent of the deletions, and the DNA fragments that were mapped are identified at the right. Notice that fragment 270, for example, failed to bind to deletions 35, 8, 10, 7, 9, 23, 24, A2, 27A, and 4D but did bind to the other deletions. The region spanned by 11q13.5 and 11q21 is missing from all the deletions that did not bind 270; 270 was thus inferred to lie within this region. Chromosome mutations often arise in cancer cells, as an example, figure shows some deletions consistently found in solid tumors. Not all the cells in a tumor show the deletion indicated, and often a mixture of different chromosome mutations can be found in one tumor. The contribution of such changes to the cancer phenotype is not understood.

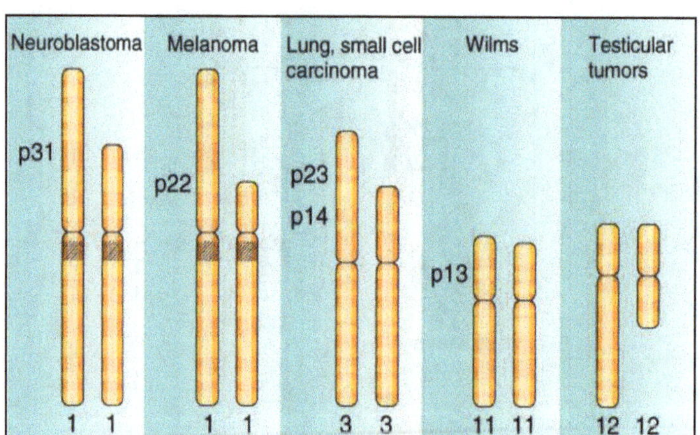

Deletions found consistently in several different types of solid tumors in humans. Band numbers indicate recurrent breakpoints.

An interesting difference between animals and plants is revealed by deletions. A male animal that is heterozygous for a deletion chromosome and a normal one produces functional sperm carrying each of the two chromosomes in approximately equal numbers. In other words, sperm seem to function to some extent regardless of their genetic content. In diploid plants, on the other hand, the pollen produced by a deletion heterozygote is of two types: (1) functional pollen carrying the normal chromosome and (2) nonfunctional (or aborted) pollen carrying the deficient homolog. Thus, pollen cells seem to be sensitive to changes in amount of chromosomal material, and this sensitivity might act to weed out deletions. The situation is somewhat different for polyploid plants, which are far more tolerant of pollen deletions. This tolerance is due to the fact that even the pollen carries several chromosome sets, and the loss of a segment in one of these sets is less crucial than it would be in a haploid pollen cell. Ovules in either diploid or polyploid plants also are quite tolerant of deletions, presumably because of the nurturing effect of the surrounding maternal tissues.

Gene Duplication

Gene duplication is the process by which a region of DNA coding for a gene is copied. Gene duplication can occur as the result of an error in recombination or through a retrotransposition event. Duplicate genes are often immune to the selective pressure under which genes normally exist. This can result in a large number of mutations accumulating in the duplicate gene code. This may render the gene non-functional or in some cases confer some benefit to the organism. There are multiple mechanisms by which gene duplication can occur.

Ectopic Recombination

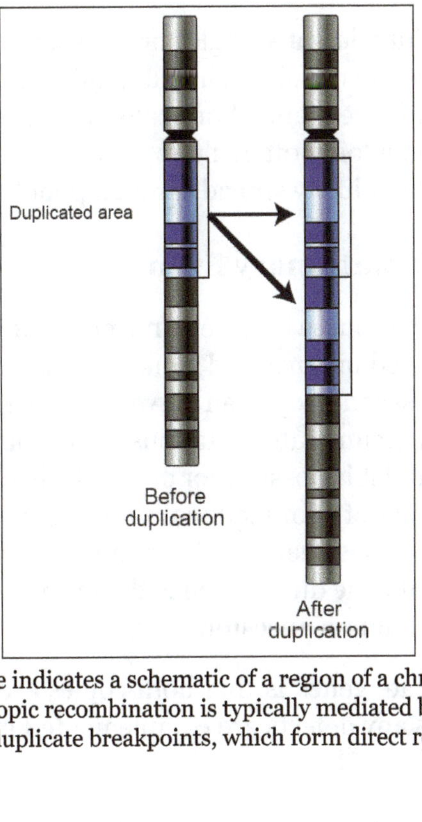

Gene Duplication: This figure indicates a schematic of a region of a chromosome before and after a duplication event. Ectopic recombination is typically mediated by sequence similarity at the duplicate breakpoints, which form direct repeats.

Duplications can arise from unequal crossing-over that occurs during meiosis between misaligned homologous chromosomes. The product of this recombination is duplication at the site of the exchange and a reciprocal deletion. Ectopic recombination is typically mediated by sequence similarity at the duplicate breakpoints, which form direct repeats. Repetitive genetic elements, such as transposable elements, offer one source of repetitive DNA that can facilitate recombination, and they are often found at duplication breakpoints in plants and mammals.

Replication Slippage

Replication slippage is an error in DNA replication, which can produce duplications of short genetic sequences. During replication, DNA polymerase begins to copy the DNA, and at some point during the replication process, the polymerase dissociates from the DNA and replication stalls. When the polymerase reattaches to the DNA strand, it aligns the replicating strand to an incorrect position and incidentally copies the same section more than once. Replication slippage is also often facilitated by repetitive sequence but requires only a few bases of similarity.

Retrotransposition

During cellular invasion by a replicating retroelement or retrovirus, viral proteins copy their genome by reverse transcribing RNA to DNA. If viral proteins attach irregularly to cellular mRNA, they can reverse-transcribe copies of genes to create retrogenes. Retrogenes usually lack intronic sequence and often contain poly A sequences that are also integrated into the genome. Many retrogenes display changes in gene regulation in comparison to their parental gene sequences, which sometimes results in novel functions.

Aneuploidy

Aneuploidy occurs when nondisjunction at a single chromosome results in an abnormal number of chromosomes. Aneuploidy is often harmful and in mammals regularly leads to spontaneous abortions. Some aneuploid individuals are viable. For example, trisomy 21 in humans leads to Down syndrome, but it is not fatal. Aneuploidy often alters gene dosage in ways that are detrimental to the organism and therefore, will not likely spread through populations.

Gene Duplication as an Evolutionary Event

Gene duplications are an essential source of genetic novelty that can lead to evolutionary innovation. Duplication creates genetic redundancy and if one copy of a gene experiences a mutation that affects its original function, the second copy can serve as a 'spare part' and continue to function correctly. Thus, duplicate genes accumulate mutations faster than a functional single-copy gene, over generations of organisms, and it is possible for one of the two copies to develop a new and different function. This is an examples of neofunctionalization. Gene duplication is believed to play a major role in evolution; this stance has been held by members of the scientific community for over 100 years. It has been argued that gene duplication is the most important evolutionary force since the emergence of the universal common ancestor.

Another possible fate for duplicate genes is that both copies are equally free to accumulate degenerative mutations, so long as any defects are complemented by the other copy. This leads to a neutral "subfunctionalization" model, in which the functionality of the original gene is distributed among the two copies. Neither gene can be lost, as both now perform important non-redundant functions, but ultimately neither is able to achieve novel functionality. Subfunctionalization can occur through neutral processes in which mutations accumulate with no detrimental or beneficial effects. However, in some cases subfunctionalization can occur with clear adaptive benefits. If an ancestral gene is pleiotropic and performs two functions, often times neither one of these two functions can be changed without affecting the other function. In this way, partitioning the ancestral functions into two separate genes can allow for adaptive specialization of subfunctions, thereby providing an adaptive benefit.

Divergence

Genetic divergence is the process in which two or more populations of an ancestral species accumulate independent genetic changes through time, often after the populations have become reproductively isolated for some period of time. In some cases, subpopulations living in ecologically distinct peripheral environments can exhibit genetic divergence from the remainder of a population, especially where the range of a population is very large. The genetic differences among divergent populations can involve silent mutations (that have no effect on the phenotype) or give rise to significant morphological and/or physiological changes. Genetic divergence will always accompany reproductive isolation, either due to novel adaptations via selection and/or due to genetic drift, and is the principal mechanism underlying speciation.

Genetic drift or allelic drift is the change in the frequency of a gene variant (allele) in a population due to random sampling. The alleles in the offspring are a sample of those in the parents, and chance has a role in determining whether a given individual survives and reproduces. A population's allele frequency is the fraction of the copies of one gene that share a particular form. Genetic drift may cause gene variants to disappear completely and thereby reduce genetic variation. When there are few copies of an allele, the effect of genetic drift is larger, and when there are many copies the effect is smaller. These changes in gene frequency can contribute to divergence.

Divergent evolution is usually a result of diffusion of the same species to different and isolated environments, which blocks the gene flow among the distinct populations allowing differentiated fixation of characteristics through genetic drift and natural selection. Divergent evolution can also be applied to molecular biology characteristics. This could apply to a pathway in two or more organisms or cell types. This can apply to genes and proteins, such as nucleotide sequences or protein sequences that are derived from two or more homologous genes. Both orthologous genes (resulting from a speciation event) and paralogous genes (resulting from gene duplication within a population) can be said to display divergent evolution.

Chromosome Abnormality

Almost every cell in our body contains 23 pairs of chromosomes, for a total of 46 chromosomes. Half of the chromosomes come from our mother, and the other half come from our father. The first 22 pairs are called autosomes. The 23rd pair consists of the sex chromosomes, X and Y. Females usually have two X chromosomes, and males usually have one X and one Y chromosome in each cell.

All of the information that the body needs to grow and develop comes from the chromosomes. Each chromosome contains thousands of genes, which make proteins that direct the body's development, growth, and chemical reactions. Many types of chromosomal abnormalities exist, but they can be categorized as either numerical or structural. Numerical abnormalities are whole chromosomes either missing from or extra to the normal pair. Structural abnormalities are when part of an individual chromosome is missing, extra, switched to another chromosome, or turned upside down.

Chromosomal abnormalities can occur as an accident when the egg or the sperm is formed or during the early developmental stages of the fetus. The age of the mother and certain environmental factors may play a role in the occurrence of genetic errors. Prenatal screening and testing can be performed to examine the chromosomes of the fetus and detect some, but not all, types of chromosomal abnormalities. Chromosomal abnormalities can have many different effects, depending on the specific abnormality. For example, an extra copy of chromosome 21 causes Down syndrome (trisomy 21). Chromosomal abnormalities can also cause miscarriage, disease, or problems in growth or development.

The most common type of chromosomal abnormality is known as aneuploidy, an abnormal chromosome number due to an extra or missing chromosome. Most people with aneuploidy have trisomy (three copies of a chromosome) instead of monosomy (single copy of a chromosome). Down syndrome is probably the most well-known example of a chromosomal aneuploidy. Besides trisomy 21, the major chromosomal aneuploidies seen in live-born babies are: trisomy 18; trisomy 13; 45, X (Turner syndrome); 47, XXY (Klinefelter syndrome); 47, XYY; and 47, XXX.

Structural chromosomal abnormalities result from breakage and incorrect rejoining of chromosomal segments. A range of structural chromosomal abnormalities result in disease. Structural rearrangements are defined as balanced if the complete chromosomal set is still present, though rearranged, and unbalanced if information is additional or missing. Unbalanced rearrangements include deletions, duplications, or insertions of a chromosomal segment. Ring chromosomes can result when a chromosome undergoes two breaks and the broken ends fuse into a circular chromosome. An isochromosome can form when an arm of the chromosome is missing and the remaining arm duplicates.

Balanced rearrangements include inverted or translocated chromosomal regions. Since the full complement of DNA material is still present, balanced chromosomal rearrangements may go undetected because they may not result in disease. A disease can arise as a result of a balanced rearrangement if the breaks in the chromosomes occur in a gene, resulting in an absent or non-functional protein, or if the fusion of chromosomal segments results in a hybrid of two genes, producing a new protein product whose function is damaging to the cell.

Pleiotropy

Pleiotropy refers to the expression of multiple traits by a single gene. These expressed traits may or may not be related. Pleitropy was first noticed by geneticist Gregor Mendel, who is known for his famous studies with pea plants. Mendel noticed that plant flower color (white or purple) was always related to the color of the leaf axil (area on a plant stem consisting of the angle between the leaf and upper part of the stem) and seed coat.

The study of pleitropic genes is important to genetics as it helps us to understand how certain traits are linked in genetic diseases. Pleitropy can be spoken of in various forms: gene pleiotropy, developmental pleiotropy, selectional pleiotropy, and antagonistic pleiotropy.

In pleiotropy, one gene controls the expression of several phenotypic traits. Phenotypes are traits that are physically expressed such as color, body shape, and height. It is often difficult to detect which traits may be the result of pleitoropy unless a mutation occurs in a gene. Because pleiotropic genes control multiple traits, a mutation in a pleiotropic gene will impact more than one trait.

Typically, traits are determined by two alleles (variant form of a gene). Specific allele combinations determine the production of proteins which drive the processes for the development of phenotypic traits. A mutation occurring in a gene alters the DNA sequence of the gene. Changing gene segment sequences most often results in non-functioning proteins. In a pleiotropic gene, all of the traits associated with the gene will be altered by the mutation.

- Gene pleiotropy, also referred to as molecular-gene pleiotropy, focuses on the number of functions of a particular gene. The functions are determined by the number of traits and biochemical factors impacted by a gene. Biochemical factors include the number of enzyme reactions catalyzed by the protein products of the gene.
- Developmental pleiotropy focuses on mutations and their influence on multiple traits. The mutation of a single gene manifests in the alteration of several different traits. Diseases involving mutational pleiotropy are characterized by deficiencies in multiple organs that impact several body systems.
- Selectional pleiotropy focuses on the number of separate fitness components affected by a gene mutation. The term fitness relates to how successful a particular organism is at transferring its genes to the next generation through sexual reproduction. This type of pleiotropy is concerned only with the impact of natural selection on traits.

Examples of Pleiotropy

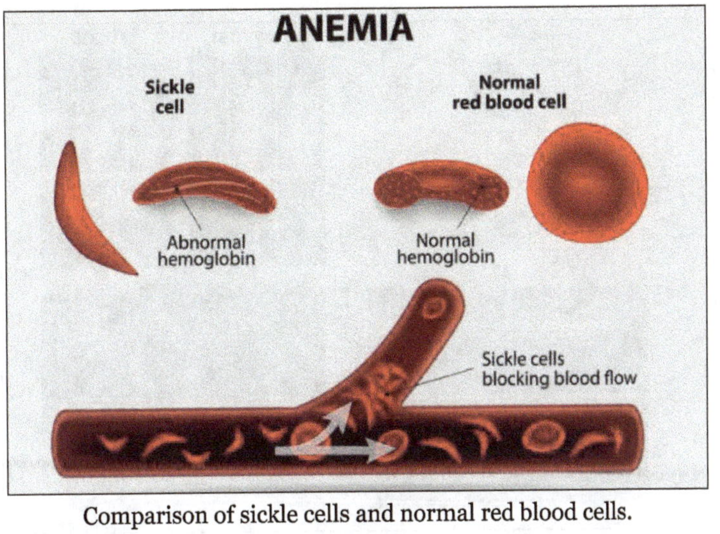

Comparison of sickle cells and normal red blood cells.

An example of pleiotropy that occurs in humans is sickle cell disease. Sickle cell disorder results from the development of abnormally shaped red blood cells. Normal red blood cells have a biconcave, disc-like shape and contain enormous amounts of a protein called hemoglobin.

Hemoglobin helps red blood cells bind to and transport oxygen to cells and tissues of the body. Sickle cell is a result of a mutation in the beta-globin gene. This mutation results in red blood cells that are sickle-shaped, which causes them to clump together and become stuck in blood vessels, blocking normal blood flow. The single mutation of the beta-globin gene results in various health complications and causes damage to multiple organs including the heart, brain, and lungs.

Phenylketonuria

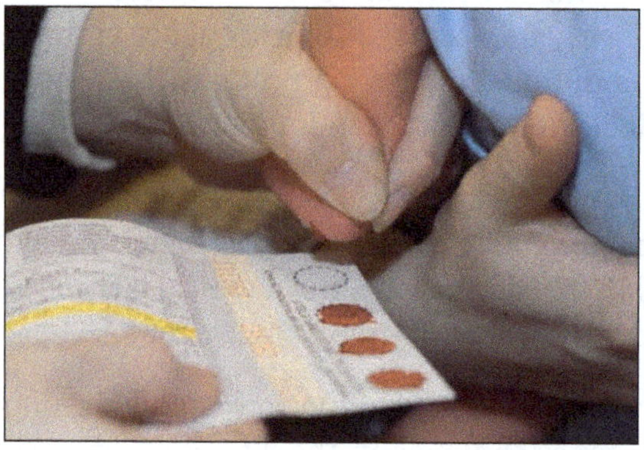

PKU Testing.

Phenylketonuria, or PKU, is another disease resulting from pleiotropy. PKU is caused by a mutation of the gene responsible for the production of an enzyme called phenylalanine hydroxylase. This enzyme breaks down the amino acid phenylalanine that we get from protein digestion. Without this enzyme, levels of the amino acid phenylalanine increase in the blood and damage the nervous system in infants. PKU disorder may result in several conditions in infants including intellectual disabilities, seizures, heart problems, and developmental delays.

Frizzled Feather Trait

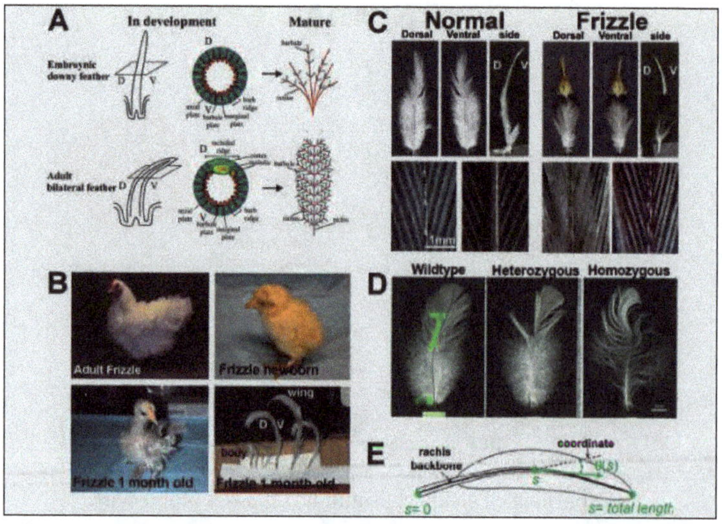

This image illustrates aspects of the frizzle chicken phenotype. Adult homozygous frizzle chicken feathers curve away from the body.

The frizzled feather trait is an example of pleiotropy seen in chickens. Chickens with this particular mutated feather gene display feathers that curl outward as opposed to lying flat. In addition to curled feathers, other pleiotropic effects include a faster metabolism and enlarged organs. The curling of the feathers leads to a loss of body heat requiring a faster basal metabolism to maintain homeostasis. Other biological changes include higher food consumption, infertility, and sexual maturation delays.

Antagonistic Pleiotropy Hypothesis

Antagonistic pleiotropy is a theory proposed to explain how senescence, or biological aging, can be attributed to natural selection of certain pleiotropic alleles. In antagonistic pleiotropy, an allele that has a negative impact on an organism can be favored by natural selection if the allele also produces advantageous effects. Antagonistically pleiotropic alleles that increase reproductive fitness early in life but promote biological aging later in life tend to be selected for by natural selection. The positive phenotypes of the pleiotropic gene are expressed early when reproductive success is high, while the negative phenotypes are expressed later in life when reproductive success is low.

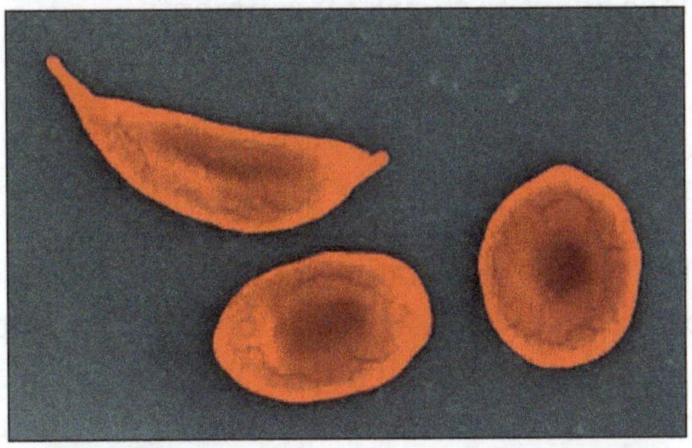

SEM of sickle cell and normal red blood cells.

Sickle cell trait is an example of antagonistic pleiotropy in that the Hb-S allele mutation of the hemoglobin gene provides advantages and disadvantages for survival. Those who are homozygous for the Hb-S allele, meaning that they have two Hb-S alleles of the hemoglobin gene, have a short life span due to the negative impact (damage to multiple body systems) of the sickle cell trait. Those who are heterozygous for the trait, meaning that they have one Hb-S allele and one normal allele of the hemoglobin gene, do not experience the same degree of negative symptoms and show resistance to malaria. The frequency of the Hb-S allele is higher in populations and regions where malaria rates are high.

Ploidy

In genetics, ploidy is the number of chromosomes occurring in the nucleus of a cell. In normal somatic (body) cells, the chromosomes exist in pairs. The condition is called diploidy. During meiosis the cell produces gametes, or germ cells, each containing half the normal or somatic number of chromosomes. This condition is called haploidy. When two germ cells (e.g., egg and sperm) unite, the diploid condition is restored. Polyploidy refers to cells the nuclei of which have three or more times the number of chromosomes found in haploid cells.

This condition frequently occurs in plants and may result from chromosome duplication without division of the cytoplasm or from the union of two diploid gametes. Polyploid animals, because they have more than the normal number of sex chromosomes, are usually sterile. Some cells have an abnormal number of chromosomes that is not a whole multiple of the haploid number. This condition, called aneuploidy, is most often caused by some error resulting in an unequal distribution of chromosomes to the daughter cells. Organisms in which aneuploidy occurs may deviate noticeably from the norm in appearance and behavior.

Haploid

Haploid is the condition of a cell having a one set of chromosomes. *Ploidy* refers to the number of copies of the genome. Humans, and many other organisms, are *diploid* organisms. This means that the majority of their lifecycle is spent with two copies of the genome in every cell. Typically, haploid cells are created for reproductive purposes. By reducing the genome to one copy, different copies can be rearranged when creating a zygote. By reducing the DNA material in the gametes to haploid, many new combinations are possible within the offspring. This increases the *genetic variation* and helps populations adapt to their environment.

Examples of Haploid

Haploid Cells in Humans

For the entirety of your life, the cells in your body are diploid, with a few exceptions. Your mother and father produced gametes, haploid cells, which came together to produce the first cell of your body. This single celled zygote replicated both copies of DNA before dividing into two identical daughter cells. The cells continued replicating and dividing until they formed a small ball, the *blastula*, which began folding and differentiating into various body parts. The cells in your body will remain diploid, as they continue replicating through *mitosis*. However, your reproductive organs will serve a special purpose. Instead of copying themselves through mitosis, certain parts of the tissues will undergo *meiosis*. Unlike mitosis, meiosis divides the *homologous chromosomes* and reduces the ploidy of the daughter cells created. These special gametes, eggs and sperm, are now the only haploid cells in your body. They are prepped to find a gamete of the opposite sex and produce a new zygote.

Haploid Drones in Insects

Many species of insect have a special sex determination system, which relies on the ploidy of the individual involved. Check out the diagram below, representing breeding systems in many bees and ants. The queen can be found in the upper-left. The queen, and all the worker bees are diploid organisms. These bees do the majority of the work in the colony, including gathering food, rearing the young, and disposing of the dead.

To the right of the queen is the haploid drone. This male insect has one simple job: carry sperm to other colonies. The queens of each colony use this sperm to fertilize their eggs, which are also haploid. Combining two haploid cells creates a diploid cell. Typically, these diploid larvae develop into average worker bees. However, if fed "royal jelly" the worker will develop into a queen. The special food activates various pathways which make the worker larger and allow her to lay eggs.

Once a hive is established, the old queen will give birth to a successor and leave the hive with many of the workers to establish a new hive. The new queen must wait to be fertilized by a haploid drone before laying new workers.

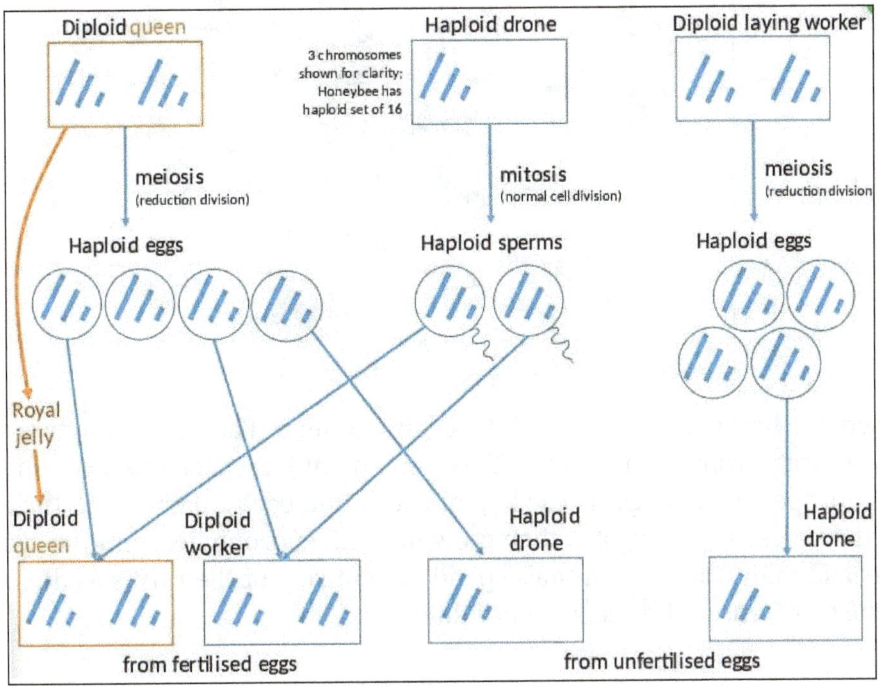

Diploid

Diploid describes a cell or nucleus which contains two copies of genetic material, or a complete set of chromosomes, paired with their homologs (chromosome carrying the same information from the other parent). By maintaining two copies of the genetic code, diploid organisms obtain an advantage by having greater genetic variation within their population, as each individual can express two alleles for each gene. Other organisms cycle between diploid and haploid lifecycles.

Examples of Diploid

Forming a Zygote

In mammals, a diploid zygote is created when two haploid gametes meet and form a single cell. This process adds the haploid DNA from each gamete into a combined diploid genome of the new zygote. Many animals reproduce using this method, although not all. The haploid stage of most animals is restricted to the single-celled gametes used for reproduction. The rest of the lifecycle is spent as a diploid, multi-celled organism.

Lifecycle of a Fern

Unlike humans and other mammals, ferns have an entire multi-celled stage of their lifecycle which is not diploid. Look at the diagram below. During the sporophyte phase, the plant is diploid. This diploid plant creates spores through meiosis, which are now haploid. The haploid cells are released into the air and travel to a new area.

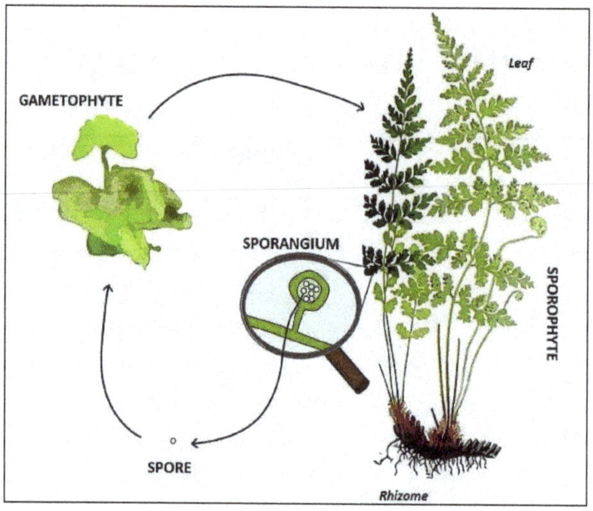

Once established, the haploid spore grows into a gametophyte. The gametophyte is an entire haploid organism, separate from the first plant. This small plant has special tissues which create gametes in the form of sperm and eggs. These haploid cells find each other and fertilize one another, creating diploid zygotes. These zygotes then grow into full sporophytes, and the cycle starts over. Where humans and many familiar animals spend the entirety of their lives as diploid organisms, many species such as ferns and insects are not that way.

Polyploidy

Polyploidy is very common in plants, especially in angiosperms. From 30% to 70% of today's angiosperms are thought to be polyploid. Species of coffee plant with 22, 44, 66, and 88 chromosomes are known. This suggests that the ancestral condition was a plant with a haploid (n) number of 11 and a diploid (2n) number of 22, from which evolved the different polyploid descendants. In fact, the chromosome content of most plant groups suggests that the basic angiosperm genome consists of the genes on 7–11 chromosomes. Domestic wheat, with its 42 chromosomes, is probably hexaploid (6n), where n (the ancestral haploid number) was 7.

Table: Some other examples of polyploidy.

Plant	Probable ancestral haploid number	Chromosome number	Ploidy level
domestic oat	7	42	6n
peanut	10	40	4n
sugar cane	10	80	8n
banana	11	22, 33	2n, 3n
white potato	12	48	4n
tobacco	12	48	4n
cotton	13	52	4n
apple	17	34, 51	2n, 3n

Polyploid plants not only have larger cells but the plants themselves are often larger. This has led to the deliberate creation of polyploid varieties of such plants as watermelons, marigolds, and snapdragons.

Origin of Polyploidy

- Polyploidy has occurred often in the evolution of plants. The process can begin if diploid (2n) gametes are formed. These can arise in at least two ways:
    - The gametes may be formed by mitosis instead of meiosis.
    - Plants, in contrast to animals, form germ cells (sperm and eggs) from somatic tissues. If the chromosome content of a precursor somatic cell has accidentally doubled (e.g., as a result of passing through S phase of the cell cycle without following up with mitosis and cytokinesis), then gametes containing 2n chromosomes are formed.
- Polyploidy also occurs naturally in certain plant tissues:
    - As the endosperm (3n) develops in corn (maize) kernels (*Zea mays*), its cells undergo successive rounds (as many as 5) of endoreplication producing nuclei that range as high as 96n.
    - When rhizobia infect the roots of their legume host, they induce the infected cells to undergo endoreplication producing cells that can become 128n (from 6 rounds of endoreplication).

Polyploidy can also be induced in the plant-breeding laboratory by treating dividing cells with colchicine. This drug disrupts microtubules and thus prevents the formation of a spindle. Consequently, the duplicated chromosomes fail to separate in mitosis. Onion cells exposed to colchicine for several days may have over 1000 chromosomes inside.

Polyploidy and Speciation

When a newly-arisen tetraploid (4n) plant tries to breed with its ancestral species (a backcross), triploid offspring are formed. These are sterile because they cannot form gametes with a balanced assortment of chromosomes. However, the tetraploid plants can breed with each other. So in one generation, a new species has been formed. Polyploidy even allows the formation of new species derived from different ancestors. In 1928, the Russian plant geneticist Karpechenko produced a new species by crossing a cabbage with a radish. Although belonging to different genera (Brassica and Raphanus respectively), both parents have a diploid number of 18. Fusion of their respective gametes (n=9) produced mostly infertile hybrids.

However, a few fertile plants were formed, probably by the spontaneous doubling of the chromosome number in somatic cells that went on to form gametes (by meiosis). Thus these contained 18 chromosomes — a complete set of both cabbage (n=9) and radish (n=9) chromosomes. Fusion of these gametes produced vigorous, fully-fertile, polyploid plants with 36 chromosomes. (They had the roots of the cabbage and the leaves of the radish.) These plants could breed with each other, but not with either the cabbage or radish ancestors, so Karpechenko had produced a new species. The process also occurs in nature. Three species in the mustard family (Brassicaceae) appear to have arisen by hybridization and polyploidy from three other ancestral species:

- B. oleracea (cabbage, broccoli, etc.) hybridized with B. nigra (black mustard) → B. carinata (Abyssinian mustard).
- B. oleracea × B. rapa (turnips) → B. napus (rutabaga).
- B. nigra × B. rapa → B. juncea (leaf mustard).

Modern wheat and perhaps some of the other plants listed in the table above have probably evolved in a similar way.

Polyploidy in Animals

Polyploidy is much rarer in animals. It is found in some insects, fishes, amphibians, and reptiles. Until recently, no polyploid mammal was known. However, the 23 September 1999 issue of Nature reported that a polyploid (tetraploid; $4n = 102$) rat has been found in Argentina. Polyploid cells are larger than diploid ones; not surprising in view of the increased amount of DNA in their nucleus. The liver cells of the Argentinian rat are larger than those of its diploid relatives, and its sperm are huge in comparison. Normal mammalian sperm heads contain some 3.3 picograms (10^{-12} g) of DNA; the sperm of the rat contains 9.2 pg. Although only one mammal is known to have all its cells polyploid, many mammals have polyploid cells in certain of their organs, e.g. the liver.

Euploidy

Euploldy is the presence of chromosome number which is the multiple of the basic chromosome set. An organism with the basic chromosome number 7, may have euploids with chromosome number 7, 14, 21, 28, 35, 42. Euploids are further of different types – monoploids, diploids and polyploids. In monoploids there is a single set of genome, in diploids there are two sets of genome and in polyploids there are more than two sets of genome.

Table: Variations involving entire sets of chromosomes.

Euploid Type	Number of homologues present for each chromosome (x)	Example
Monoploide or haploid	One (x)	A B C
Diploid	Two (2x)	AA BB CC
Polyploid	More then 2	
Triploid	Three 3(x)	AAA BBB CCC
Tetraploid	Four (4x)	AAAA BBBB CCCC
Poentaploid	Five (5x)	AAAAA BBBBB CCCCC
Hexaploid	Six (6x)	AAAAAA BBBBBB CCCCCC
Septaploid	Seven (7x)	AAAAAAA BBBBBBB CCCCCCC
Octaploid, etc.	Eight (8x), etc.	

Symbols of chromosome number:

- 2n = Somatic chromosome number of a diploid or polyploid species.
- n = Gametic chromosome number of a diploid or polyploid species.
- X = Basic chromosome number or genomic number.
- In a diploid species with chromosome number 2n = 14, n = 7 as well as x = 7. But in a polyploid (hexaploid) species with chromosome number 2n = 6x = 42, n = 21 but X = 7.

Types of Euploidy

Monoploidy and Haploidy

Monoploid individuals have single basic set of chromosome, e.g., in barley 2n = x = 7 (haploid of a diploid species). Haploids are individuals with chromosome number half of the somatic number, e.g., in wheat (2n = 3x = 21). In diploid species, the chromosome number of monoploids and haploids are same, but in polyploid species the chromosome number of monoploids and haploids are different. In polyploid wheat (6x = 42), the haploid is 3x = 21 and manoploid is x = 7. In flowering plants, the diplophase or the sporophytic phase dominates; the haplophase or gametophytic phase is normally limited to the pollen grains and the embryo sacs. In exceptional cases, plants may arise that are entirely haploid. With regard to all their parts, the haploids are smaller and often display poor vigour.

Meiosis

Meiosis in a true haploid is of course very irregular and the chromosomes are distributed at random, some of them move to one pole, others to the opposite pole. Moreover, there is often chromosome elimination; i.e., single chromosomes are eliminated in the cytoplasm and are not included in the pollen grains or embryo sacs. In most cases, pollen grains and embryo sacs will receive an incomplete or unbalanced chromosome constitution, which will have a lethal effect. Thus, the haploids are completely or almost completely sterile.

Types

Haploids are of different types. Polyhaploids are haploids obtained from polyploid species, for instance, wheat group, which is hexaploid with 2n = 42, contains more than one genome. If the genomes are homologous or partially homologous, bivalent formation and good fertility in the polyhaploid may be the consequence. Aneuhaploids are also available and of two types — disomic haploid (haploid obtained from a tetrasomic) and nullisomic haploid (deficient of one chromosome than euhaploid).

Origin

Haploids may originate spontaneously but, as a rule, in only very low frequencies. The haploids may be induced through:

- Delayed pollination and the egg dividing without fertilization.

- X-raying of pollen.
- Pollinating with an incompatible species, for instance, Solanum nigrum is pollinated with pollen from S. luteum, haploids of S. nigrum may arise. These haploids are derived from unfertilized egg ceils of S. nigrum.
- A fourth method to obtain haploids and polyembryony is by the twin method.
- Lastly, the most important method for induction of haploidy is by culturing pollen grain in vitro as done in case of species of Datura, Oryza and some other agricultural crops.

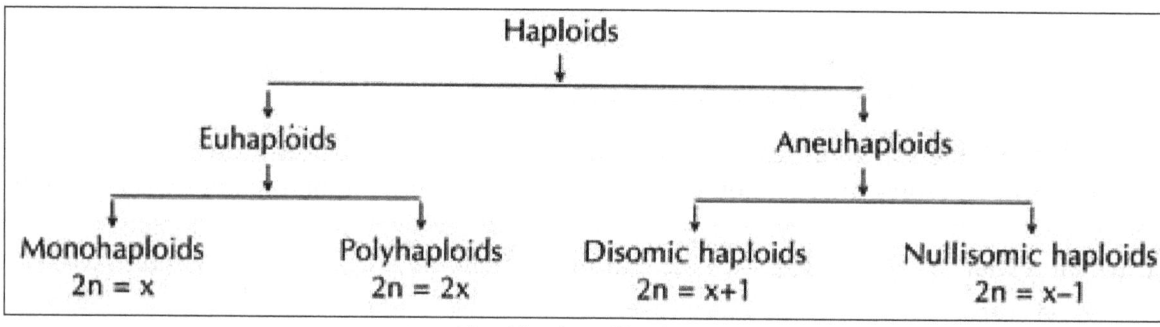

Classification of haploids.

Aneuploidy

Aneuploidy is the second major category of chromosome mutations in which chromosome number is abnormal. An aneuploid is an individual organism whose chromosome number differs from the wild type by part of a chromosome set. Generally, the aneuploid chromosome set differs from wild type by only one or a small number of chromosomes. Aneuploids can have a chromosome number either greater or smaller than that of the wild type. Aneuploid nomenclature is based on the number of copies of the specific chromosome in the aneuploid state. For example, the aneuploid condition $2n - 1$ is called monosomic (meaning "one chromosome") because only one copy of some specific chromosome is present instead of the usual two found in its diploid progenitor. The aneuploid $2n + 1$ is called trisomic, $2n - 2$ is nullisomic, and $n + 1$ is disomic.

Nullisomics (2n – 2)

Although nullisomy is a lethal condition in diploids, an organism such as bread wheat, which behaves meiotically like a diploid although it is a hexaploid, can tolerate nullisomy. The four homoeologous chromosomes apparently compensate for a missing pair of homologs. In fact, all the possible 21 bread wheat nullisomics have been produced; they are illustrated in figure. Their appearances differ from the normal hexaploids; furthermore, most of the nullisomics grow less vigorously.

Monosomics (2n – 1)

Monosomic chromosome complements are generally deleterious for two main reasons. First, the missing chromosome perturbs the overall gene balance in the chromosome set. Second, having a chromosome missing allows any deleterious recessive allele on the single chromosome to be hemizygous and thus to be directly expressed phenotypically. Notice that these are the same effects produced by deletions.

Nondisjunction in mitosis or meiosis is the cause of most aneuploids. Disjunction is the normal separation of homologous chromosomes or chromatids to opposite poles at nuclear division. Nondisjunction is a failure of this disjoining process, and two chromosomes (or chromatids) go to one pole and none to the other. Nondisjunction occurs spontaneously; it is another example of a chance failure of a basic cellular process. In meiotic nondisjunction, the chromosomes may fail to disjoin at either the first or second division. Either way, $n + 1$ and $n - 1$ gametes are produced. If an $n - 1$ gamete is fertilized by an n gamete, a monosomic ($2n - 1$) zygote is produced. The fusion of an $n + 1$ and an n gamete yields a trisomic $2n + 1$.

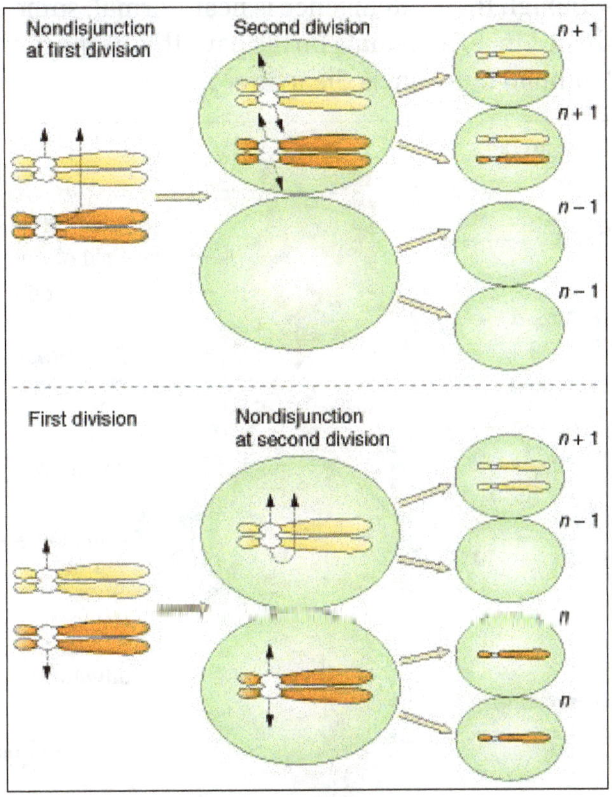

The origin of aneuploid gametes by nondisjunction at the first or second meiotic division.

The precise molecular processes that fail in nondisjunction are not known, but, in experimental systems, the frequency of nondisjunction can be increased by interference with microtubule action. It appears that disjunction is more likely to go awry in meiosis I. This likelihood may not be surprising, because normal anaphase I disjunction requires that proper homologous associations be maintained during prophase I and metaphase I. In contrast, proper disjunction at anaphase II or at mitosis requires that the centromere splits properly but does not require nearly as elaborate a process during prophase and metaphase.

Meiosis I nondisjunction can be viewed as the failure to form or maintain a tetrad (a group of four chromatids) until anaphase I. Crossovers are implicit in this process normally. In most organisms, the amount of crossing-over is sufficient to ensure that all tetrads will have at least one exchange per meiosis. In *Drosophila,* many of the nondisjunctional chromosomes in newly arising disomic gametes are nonrecombinant, with one nondisjunctional homolog carrying the markers of one input chromosome and the other homolog carrying the markers of the other chromosome. Similar observations have been made in human trisomies. In addition, in several different experimental organisms, mutations that interfere with recombination have the effect of massively increasing the frequency of meiosis I nondisjunction. This effect points to an important role of crossing-over in maintaining chromosomal associations in the tetrad; in the absence of these associations, chromosomes are vulnerable to anaphase I nondisjunction.

In humans, a sex-chromosome monosomic complement of 44 autosomes + 1 X produces a phenotype known as *Turner syndrome*. Affected people have a characteristic, easily recognizable phenotype: they are sterile females, are short in stature, and often have a web of skin extending between the neck and shoulders. Although their intelligence is near normal, some of their specific cognitive functions are defective. About 1 in 5000 female births have this monosomic chromosomal complement. Monosomics for all human autosomes die in utero.

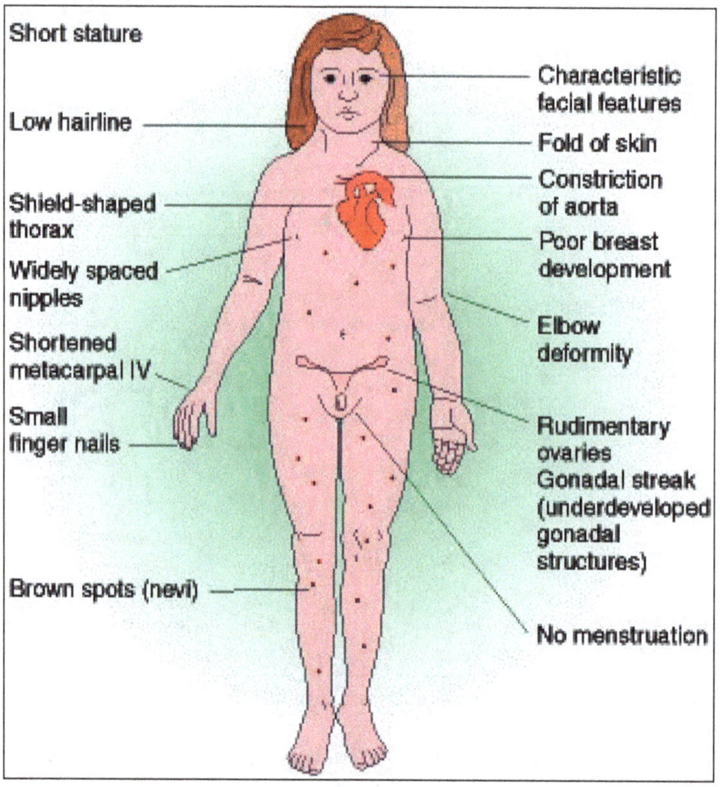

Characteristics of Turner syndrome, which results from having a single X chromosome (XO).

Geneticists have used viable plant nullisomics and monosomics to identify the chromosomes that carry the loci of newly found recessive mutant alleles. For example, a geneticist may obtain different monosomic lines, each of which lacks a different chromosome. Homozygotes for the new mutant allele are crossed with each monosomic line, and the progeny of each cross are inspected for expression of the recessive phenotype. The phenotype appears in some of the progeny of the parent that is monosomic for the locus-bearing chromosome and thus identifies it. Figure shows that this test works because meiosis in the monosomic parent produces some gametes that lack the chromosome bearing the locus. When one of these gametes forms a zygote, the single chromosome contributed by the other homozygous recessive parent determines the phenotype.

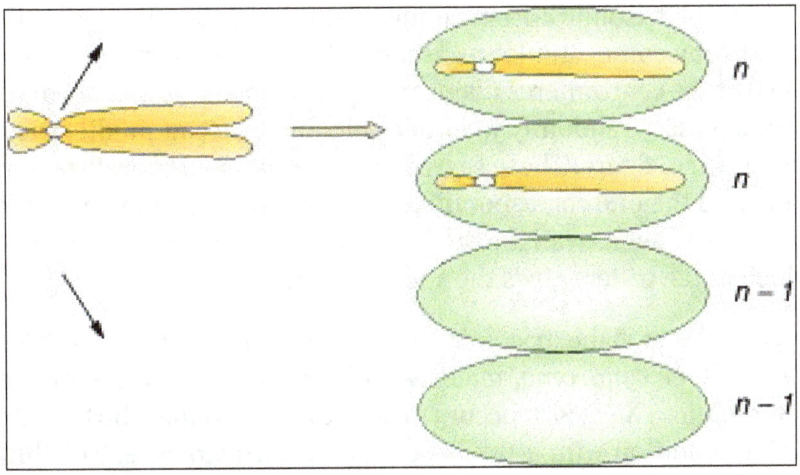

Meiosis in which the chromosome of interest is monosomic. Two of the resulting gametes are haploid (n); the other two gametes contain a set lacking a chromosome ($n - 1$).

Genetic analysis of humans occasionally reveals a similar unmasking of a recessive phenotype by an $n - 1$ gamete. For example, two people whose vision is normal may produce a daughter who has *Turner syndrome* and who is also red-green colorblind. This coincidence is interpreted as follows. First, because the father is not colorblind, the mother must be heterozygous for the recessive allele and must have passed this allele on to the Turner daughter. Nondisjunction must have occurred in the father, resulting in an $n - 1$ sperm, which combined with an egg bearing the colorblind allele from the mother.

Trisomics (2n + 1)

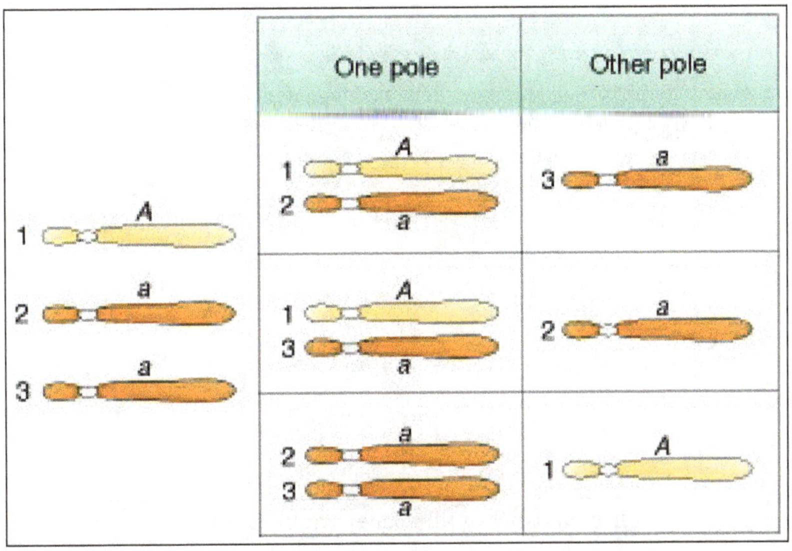

Genotypes of the meiotic products of an A/a/a trisomic. Three segregations are equally likely.

The trisomic condition also is one of chromosomal imbalance and can result in abnormality or death. However, there are many examples of viable trisomics. Furthermore, trisomics can be fertile. When cells from some trisomic organisms are observed under the microscope at the time of meiotic chromosome pairing, the trisomic chromosomes are seen to form a trivalent, an associated group of three, whereas the other chromosomes form regular bivalents. What genetic ratios might we expect for genes on the trisomic chromosome? Let us consider a gene, *A*, that is close to the centromere on that chromosome, and let us assume that the genotype is *A*/*a*/*a*. Furthermore, if we postulate that two of the centromeres disjoin to opposite poles as in a normal bivalent and that the other chromosome passes randomly to either pole, then we can predict the three equally frequent segregations shown in figure. These segregations result in an overall gametic ratio of 1*A*:2*A*/*a*:2*a*:1*a*/*a*. This ratio and the one corresponding to a trisomic of genotype *A*/*A*/*a* are observed in practice. If a trisomic tester set is available, then a new mutation can be located to a chromosome by determining which of the testers gives the special ratio.

There are several examples of viable trisomics in humans. The combination XXY (1 in 1000 male births) results in *Klinefelter syndrome,* males with lanky builds who are mentally retarded and sterile. Another combination, XYY, also occurs in about 1 in 1000 male births. Attempts have been made to link the XYY condition with a predisposition toward violence. This linkage is still hotly debated, although it is now clear that an XYY condition in no way guarantees such behavior. Nevertheless, several enterprising lawyers have attempted to use the XYY genotype as grounds for acquittal or compassion in crimes of violence. The XYY males are usually fertile. Their meioses are of the XY type; the extra Y is not transmitted, and their gametes contain either X or Y, never YY or XY.

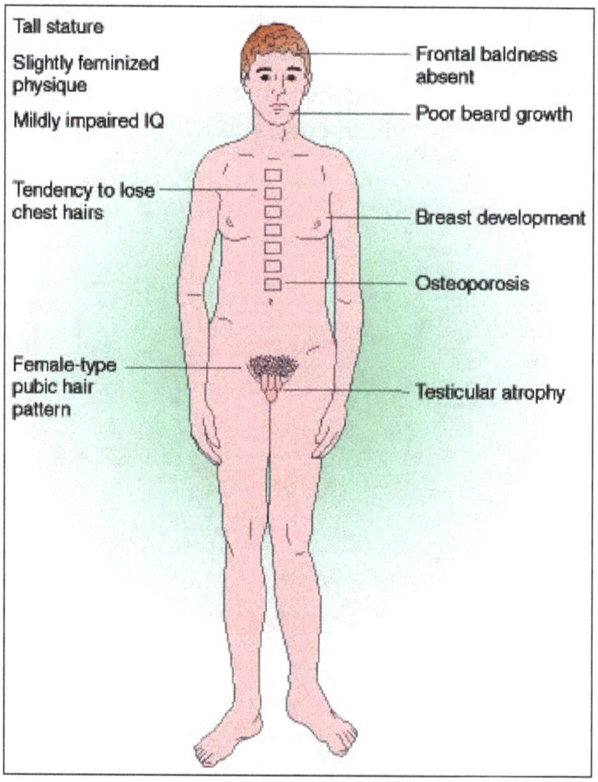

Characteristics of Klinefelter syndrome (XXY).

The most common type of viable human aneuploid is *Down syndrome*, occurring at a frequency of about 0.15 percent of all live births. However, by far the most common type of Down syndrome is trisomy 21, caused by nondisjunction of chromosome 21 in a parent who is chromosomally normal. Like any mechanism, chromosome disjunction is error prone and sometimes produces aneuploid gametes. In this type of Down syndrome, there is no family history of aneuploidy, unlike the translocation type.

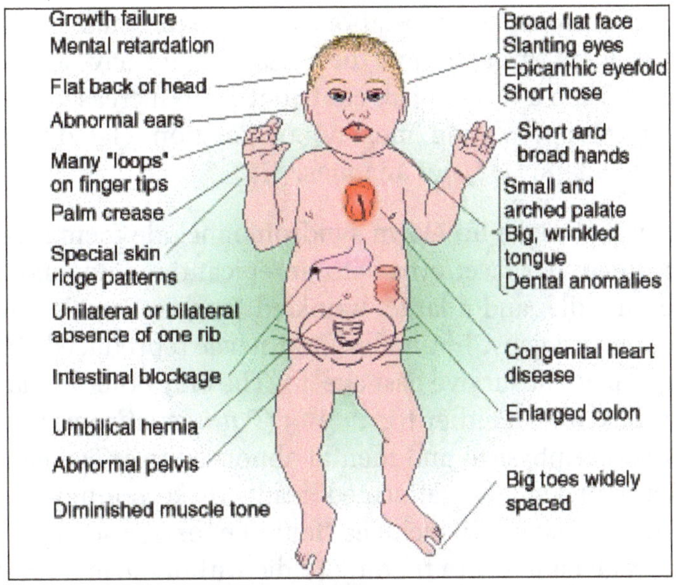

Characteristics of Down syndrome (trisomy 21). Diagrammatic representation of the syndrome in an infant.

Down syndrome is related to maternal age; older mothers run a greatly elevated risk of having Down-syndrome children. For this reason, fetal chromosome analysis (by amniocentesis or by chorionic villus sampling) is now recommended for older mothers. A less pronounced paternal-age effect also has been demonstrated.

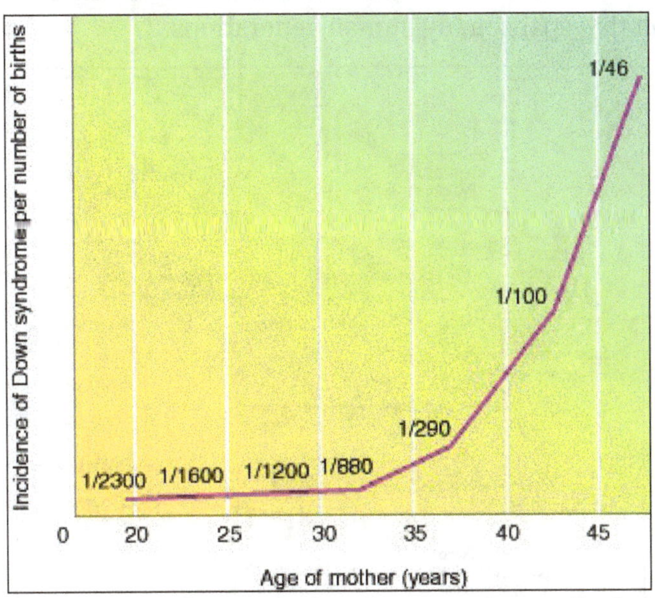

Maternal age and the production of Down-syndrome offspring.

Even though the maternal-age effect has been known for many years, its cause is still not known. Nonetheless, there are some interesting biological correlations. It is possible that one aspect of the strong maternal-age effect on nondisjunction is an age-dependent decrease in the probability of keeping the chromosomal tetrad together during prophase I of meiosis. Meiotic arrest of oocytes (female meiocytes) in late prophase I is a common phenomenon in many animals. In female humans, all oocytes are arrested at diplotene before birth. Meiosis continues only at menstruation, which means that proper chromosome associations in the tetrad must be maintained for decades. If we speculate that, by accident through time, these associations have an increasing probability of breaking down, we can envision a mechanism contributing to increased maternal nondisjunction with age. Consistent with this speculation, most nondisjunction related to the maternal-age effect is due to nondisjunction at anaphase I, not anaphase II.

The multiple phenotypes that make up Down syndrome include mental retardation, with an IQ in the 20 to 50 range; broad, flat face; eyes with an epicanthic fold; short stature; short hands with a crease across the middle; and a large, wrinkled tongue. Females may be fertile and may produce normal or trisomic progeny, but males have never reproduced. Mean life expectancy is about 17 years, and only 8 percent survive past age 40. The only other human autosomal trisomics to survive to birth are afflicted with either trisomy 13 *(Patau syndrome)* or trisomy 18 *(Edwards syndrome)*. Both show severe physical and mental abnormalities. The generalized phenotype of trisomy 13 includes a harelip; a small, malformed head; "rockerbottom" feet; and a mean life expectancy of 130 days. That of trisomy 18 includes "faunlike" ears, a small jaw, a narrow pelvis, and rockerbottom feet; almost all babies with trisomy 18 die within the first few weeks after birth.

Chromosome mutation in general plays a prominent role in determining genetic ill health in humans. Figure summarizes the surprisingly high levels of various chromosome abnormalities at different human developmental stages. In fact, the incidence of chromosome mutations ranks close to that of gene mutations in human live births. This fact is particularly surprising when we realize that virtually all the chromosome mutations listed in table arise anew with each generation. In contrast, gene mutations owe their level of incidence to a complex interplay of mutation rates and environmental selection that spans many human generations.

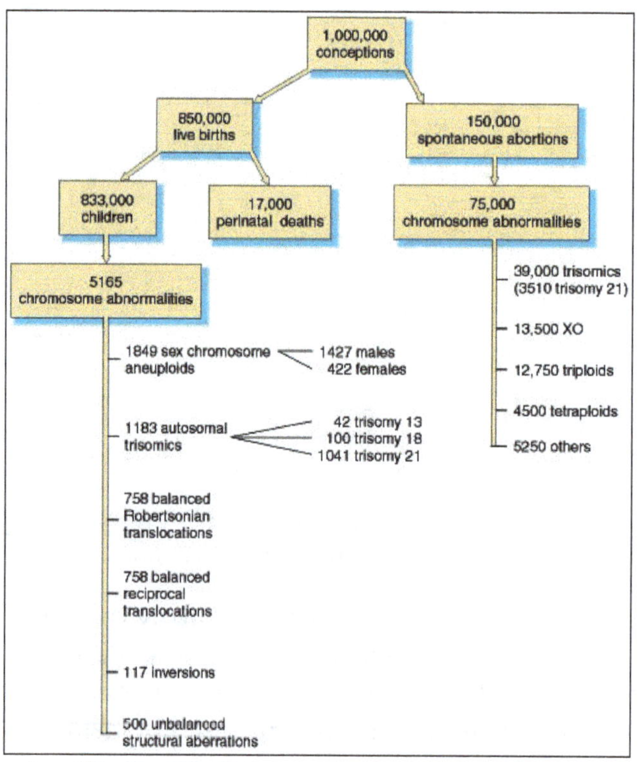

The fate of a million implanted human zygotes. (Robertsonian translocations are due to fusion or dissociation of centromeres.)

Table: Relative incidence of human ill health due to gene mutation and to chromosome mutation.

Type of mutation	Percentage of live births
Gene mutation	
Autosomal dominant	0.90
Autosomal recessive	0.25
X linked	0.05
Total gene mutation:	1.20
Chromosome mutation	
Autosomal trisomies (mainly Down syndrome)	0.14
Other unbalanced autosomal aberrations	0.06
Balanced autosomal aberrations	0.19
Sex chromosomes	
XYY, XXY, and other ♂♂	0.17
XO, XXX, and other ♀♀	0.05
Total chromosome mutation:	0.61

When the frequencies of various chromosome mutations in live births are compared with the corresponding frequencies found in spontaneous abortions, it becomes clear that the chromosome mutations that we know about as clinical abnormalities are just the tip of an iceberg of chromosome mutations. First, we see that many more types of abnormalities are produced than survive to birth; for example, trisomies of chromosome 2, 16, and 22 are relatively common in abortuses but never survive to birth. Second, the specific aberrations that survive are part of a much larger number that do not survive; for example, Down syndrome (trisomy 21) is produced at almost 20 times the frequency in live births. The comparison is even more striking for Turner syndrome (XO). An estimated minimum of 10 percent of conceptions have a major chromosome abnormality; our reproductive success depends on the natural weeding-out process that eliminates most of these abnormalities before birth. Incidentally, no evidence suggests that these aberrations are produced by environmental insult to our reproductive systems or that the frequency of the aberrations is increasing.

Table: Number and type of chromosomal abnormalities among spontaneous abortions and live births in 100,000 pregnancies.

	100,000 Pregnancies	
	15,000 spontaneous abortions 7,500 chromosomally abnormal	85,000 live births 550 chromosomally abnormal
Trisomy		
1	0	0
2	159	0
3	53	0
4	95	0

5	0	0
6–12	561	0
13	128	17
14	275	0
15	318	0
16	1229	0
17	10	0
18	223	13
19–20	52	0
21	350	113
22	424	0
Sex chromosomes		
XYY	4	46
XXY	4	44
XO	1350	8
XXX	21	44
Translocations		
Balanced	14	164
Unbalanced	225	52
Polyploid		
Triploid	1275	0
Tetraploid	450	0
Other (mosaics, etc.)	280	49
Total	7500	550

Disomics (n + 1)

A disomic is an aberration of a haploid organism. In fungi, they can result from meiotic nondisjunction. In the fungus *Neurospora* (a haploid), an $n - 1$ meiotic product aborts and does not darken like a normal ascospore; so we may detect MI and MII nondisjunctions by observing asci with 4:4 and 6:2 ratios of normal to aborted spores, respectively, as shown here.

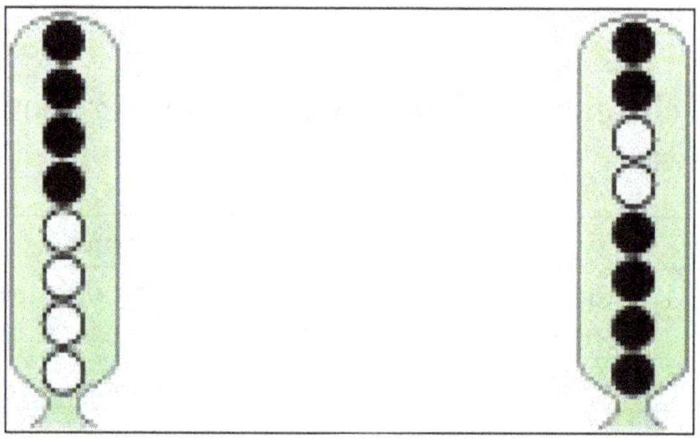

These diagrams correspond exactly to the outcomes of the chromosomal events shown in figure. In these organisms, the disomic ($n + 1$) meiotic product becomes a disomic strain directly. The abortion patterns themselves are diagnostic for the presence of disomics in the asci. Another way of detecting disomics in fungi is to cross two strains with homologous chromosomes bearing multiple auxotrophic mutations; for example:

$$\frac{\text{ade his}^+ \text{ nic ala}^+ \text{leu arg}^+}{\quad}$$

$$\times$$

$$\frac{\text{ade}^+ \text{ his nic}^+ \text{ala leu}^+ \text{ arg}}{\quad}$$

From such a cross, large populations of ascospores are plated onto minimal medium. Only ascospores of genotype + + + + + + can grow and form colonies. Most of these colonies are found to be disomics and not multiple crossover types.

Somatic Aneuploids

Aneuploid cells can arise spontaneously in somatic tissue or in cell culture. In such cases, the initial result is a genetic mosaic of cell types. Human sexual mosaics individuals whose bodies are a mixture of male and female tissue are good examples. One type of sexual mosaic, (XO)(XYY), can be explained by postulating an XY zygote in which the Y chromatids fail to disjoin at an early mitotic division, so both go to one pole:

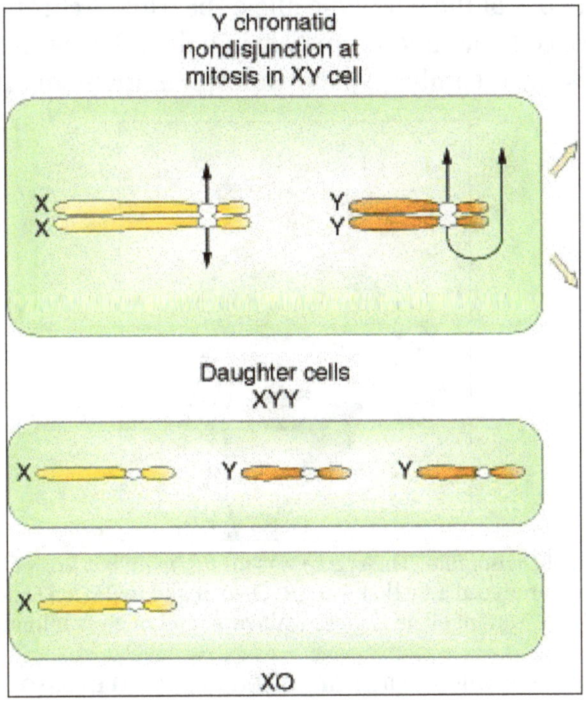

The phenotypic sex of such individuals depends on where the male and female sectors end up in the body. In the type of nondisjunction being considered, nondisjunction at a later mitotic division would produce a three-way mosaic (XY)(XO)(XYY), which contains a clone of normal male cells. Other sexual mosaics have different explanations; as examples, XO/XY is probably due to early X-chromosome loss in a male zygote, and (XX)(XY) is probably the result of a double fertilization (fused twins). In general, sexual mosaics are called *gynandromorphs*.

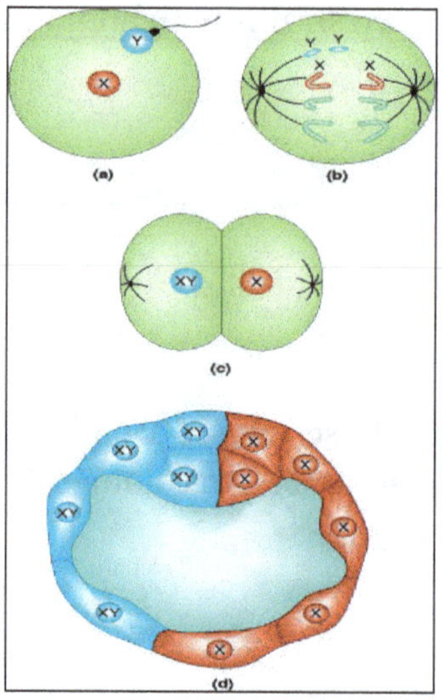

Origin of a human sexual mosaic (XY)(XO) by Y chromosome loss at the first mitotic division of the zygote. (a) Fertilization. (b) Chromosome loss. (c) Resulting male and female cells. (d) Mosaic blastocyst.

Geneticists working with many species of experimental animals occasionally find gynandromorphs among their stocks. A classic example is the *Drosophila* gynandromorph shown in figure. In this case, the zygote started out as a female heterozygous for two X-linked genes, white eye and miniature wing ($w^+ m^+/w\ m$). Loss of the wild-type allele–bearing X chromosome at the first mitotic division resulted in the two cell lines and ultimately in a fly differing from one side to the other in sex, eye color, and size of wing. A similar gynandromorph in the Io moth is shown in.

A bilateral gynandromorph of *Drosophila*. The zygote was $w^+ m^+/w\ m$, but loss of the $w^+ m^+$ chromosome in the first mitotic division produced a fly that was 1/2 O/$w\ m$ and male *(left)* and 1/2$w^+ m^+/w\ m$ and female *(right)*. Mutant allele w gives white eye, and m gives miniature wing.

Somatic aneuploidy and its resulting mosaics are often observed in association with cancer. People suffering from chronic myeloid leukemia (CML), a cancer of the white blood cells, frequently harbor cells containing the so-called Philadelphia chromosome. This chromosome was once thought to represent an aneuploid condition, but it is now known to be a translocation product in which part of the long arm of chromosome 22 attaches to the long arm of chromosome 9. However, CML patients often show aneuploidy in addition to the Philadelphia chromosome. In one study of 67 people with CML, 33 proved to have an extra Philadelphia chromosome and the remainder had various aneuploidies; the most common aneuploidy was trisomy for the long arm of chromosome

17, which was detected in 28 people. Of 58 people with acute myeloid leukemia, 21 were shown to have aneuploidy for chromosome 8; 16 for chromosome 9; and 10 for chromosome 21. In another study of 15 patients with intestinal tumors, 12 had cells with abnormal chromosomes, at least some with trisomy for chromosome 8, 13, 15, 17, or 21. Such studies merely established correlations, and it is not clear whether the abnormalities are best thought of as a cause or as an effect of cancer.

Natural Selection

Natural selection is the process that results in the adaptation of an organism to its environment by means of selectively reproducing changes in its genotype, or genetic constitution. In natural selection, those variations in the genotype (the entire complex of genes inherited from both parents) that increase an organism's chances of survival and procreation are preserved and multiplied from generation to generation at the expense of less advantageous variations. Evolution often occurs as a consequence of this process. Natural selection may arise from differences in survival, in fertility, in rate of development, in mating success, or in any other aspect of the life cycle. All such differences result in natural selection to the extent that they affect the number of progeny an organism leaves.

Gene frequencies tend to remain constant from generation to generation when disturbing factors are not present. Factors that disturb the natural equilibrium of gene frequencies include mutation, migration (or gene flow), random genetic drift, and natural selection. A mutation is a spontaneous change in the gene frequency that takes place in a population and occurs at a low rate. Migration is a local change in gene frequency when an individual moves from one population to another and then interbreeds. Random genetic drift is a change that takes place from one generation to another by a process of pure chance. Mutation, migration, and genetic drift alter gene frequencies without regard to whether such changes increase or decrease the likelihood of an organism surviving and reproducing in its environment. They are all random processes.

Natural selection moderates the disorganizing effects of these processes because it multiplies the incidence of beneficial mutations over the generations and eliminates harmful ones, since their carriers leave few or no descendants. Natural selection enhances the preservation of a group of organisms that are best adjusted to the physical and biological conditions of their environment and may also result in their improvement in some cases. Some characteristics, such as the male peacock's tail, actually decrease the individual organism's chance of survival. To explain such anomalies, Darwin posed a theory of "sexual selection." In contrast to features that results from natural selection, a structure produced by sexual selection results in an advantage in the competition for mates.

Genetic Variation

Genetic variation refers to diversity in gene frequencies. Genetic variation can refer to differences between individuals or to differences between populations. Mutation is the ultimate source of genetic variation, but mechanisms such as sexual reproduction and genetic drift contribute to it as well.

Types of Variation

Because a great deal of genetics concerns the analysis of variants, it is important to understand the types of variation found in populations. A useful classification is into discontinuous and continuous variation. Allelic variation contributes to both.

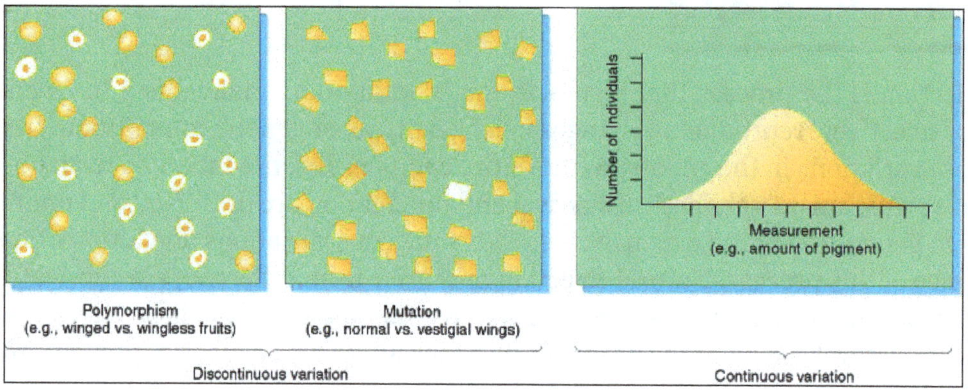

Discontinuous and continuous variation in natural populations. In populations showing discontinuous variation for a particular character, each member possesses one of several discrete alternatives. For example, in the left-hand panel, a population of plants manifests two distinct, common fruit types: winged and wingless. Such variation is called a *polymorphism.* Sometimes most of the population is of one kind, with mutation providing only the occasional alternative, as in the vestigial wing type in the fruit fly, *Drosophila.* Discontinuous variants are often determined by the alleles of a single gene. Continuous variation, on the other hand, does not show such discrete alternatives; a character may be found in phenotypic gradations in a population. There may be no genetic basis – that is, all variation may be environmentally caused – or genes may play a role, often several or many genes.

Most of the research in genetics in the past century has been on discontinuous variation because it is a simpler type of variation, and it is easier to analyze. In discontinuous variation, a character is found in a population in two or more distinct and separate forms called phenotypes. Such alternative phenotypes are often found to be encoded by the alleles of one gene. A good example is albinism in humans, which concerns phenotypes of the character of skin pigmentation. In most people, the cells of the skin can make a dark brown or black pigment called melanin, the substance that gives our skin its color ranging from tan color in people of European ancestry to brown or black in those of tropical and subtropical ancestry. Although always rare, albinos are found in all races; they have a totally pigmentless skin and hair. The difference between pigmented and unpigmented is caused by two alleles of a gene taking part in melanin synthesis.

The alleles of a gene are conventionally designated by letters. The allele that codes for the ability to make melanin is called A and the allele that codes for the inability to make melanin (resulting in albinism) is designated a to show that they are related. The allelic constitution of an organism is its genotype, which is the hereditary underpinning of the phenotype. Because humans have two sets of chromosomes in each cell, genotypes can be either A/A, A/a, or a/a (the slash shows that they are a pair). The phenotype of A/A is pigmented, a/a is albino, and A/a is pigmented. The *ability* to make pigment is expressed over *inability*. Although allelic differences cause phenotypic differences such as pigmented and albino, this does not mean that only one gene affects skin color. It is known that there are several. However, the difference between pigmented, of whatever shade, and albino is caused by the difference at one gene; the state of all the other pigment genes is irrelevant.

In discontinuous variation, there is a predictable one-to-one relation between genotype and phenotype under most conditions. In other words, the two phenotypes (and their underlying genotypes) can almost always be distinguished. In the albinism example, the A allele always allows some pigment formation, whereas the white allele always results in albinism when homozygous. For this reason, discontinuous variation has been successfully used by geneticists to identify the underlying alleles and their role in cellular functions.

Geneticists distinguish two categories of discontinuous variation on the basis of simple allelic differences. In a natural population, the existence of two or more common discontinuous variants is called polymorphism and an example is shown in figure below. The various forms are called morphs. It is often found that morphs are determined by the alleles of a single gene. Why do populations show genetic polymorphism? Special types of natural selection can explain a few cases, but, in other cases, the morphs seem to be selectively neutral.

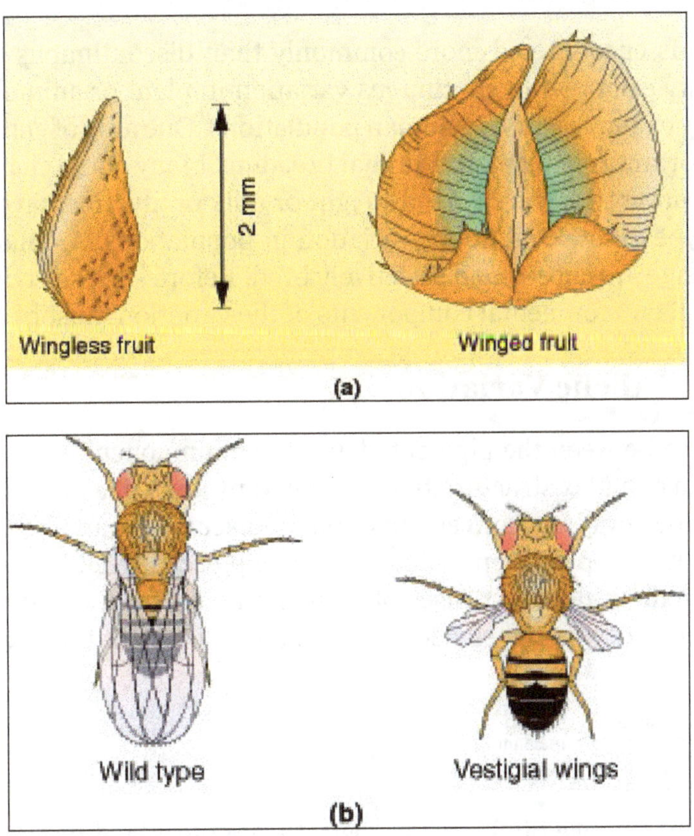

Figure shows: A dimorphism. (a) The fruits of two different forms of *Plectritis congesta,* the sea blush. Any one plant has either all wingless or all winged fruits. In every other way, the plants are identical. (b) A *Drosophila* mutant with abnormal wings and a normal fly (wild type) for comparison. In both cases, the two phenotypes are caused by the alleles of one gene.

Rare, exceptional discontinuous variants are called mutants, whereas the more common “normal” companion phenotype is called the wild type. Figure shows an example of a mutant phenotype. Again, in many cases, the wild-type and mutant phenotypes are determined by the alleles of one gene. Mutants can occur spontaneously in nature (for example, albinos) or they can be obtained after treatment with mutagenic chemicals or radiations. Geneticists regularly induce mutations artificially to carry out genetic analysis because mutations that affect some specific biological function under study identify the various genes that interact in that function. Note that polymorphisms originally arise as mutations, but somehow the mutant allele becomes common.

Continuous variation of a character shows an unbroken range of phenotypes in the population. Measurable characters such as height, weight, and color intensity are good examples of such variation. Intermediate phenotypes are generally more common than extreme phenotypes and, when phenotypic frequencies are plotted as a graph, a bell-shaped distribution is observed. In some such distributions, all the variation is environmental and has no genetic basis at all. In other cases, there is a genetic component caused by allelic variation of one or many genes. In most cases, there is both genetic and environmental variation. In continuous distributions, there is no one-to-one correspondence of genotype and phenotype. For this reason, little is known about the types of genes underlying continuous variation, and only recently have techniques become available for identifying and characterizing them.

Continuous variation is encountered more commonly than discontinuous variation in everyday life. We can all identify examples of continuous variation in plant or animal populations that we have observed – many examples exist in human populations. One area of genetics in which continuous variation is important is in plant and animal breeding. Many of the characters that are under selection in breeding programs, such as seed weight or milk production have complex determination, and the phenotypes show continuous variation in populations. Animals or plants from one extreme end of the range are chosen and selectively bred. Before such selection is undertaken, the sizes of the genetic and environmental components of the variation must be known.

Molecular Basis of Allelic Variation

Consider the difference between the pigmented and the albino phenotypes in humans. The dark pigment melanin has a complex structure that is the end product of a biochemical synthetic pathway. Each step in the pathway is a conversion of one molecule into another, with the progressive formation of melanin in a step-by-step manner. Each step is catalyzed by a separate enzyme protein encoded by a specific gene. Most cases of albinism result from changes in one of these enzymes – tyrosinase. The enzyme tyrosinase catalyzes the last step of the pathway, the conversion of tyrosine into melanin.

$$\rightarrow \underset{\text{(not a pigment)}}{\text{tyrosine}} \xrightarrow{\text{tyrosinase}} \underset{\text{(pigment)}}{\text{melanin}}$$

To perform this task, tyrosinase binds to its substrate, a molecule of tyrosine, and facilitates the molecular changes necessary to produce the pigment melanin. There is a specific “lock-and-key” fit between tyrosine and the active site of the enzyme. The active site is a pocket formed by several crucial amino acids in the polypeptide. If the DNA of the tyrosinase-encoding gene changes in such a way that one of these crucial amino acids is replaced by another amino acid or lost, then there are several possible consequences. First, the enzyme might still be able to perform its functions but in a less efficient manner. Such a change may have only a small effect at the phenotypic level, so small as to be difficult to observe, but it might lead to a reduction in the amount of melanin formed and, consequently, a lighter skin coloration. Note that the protein is still present more or less intact, but its ability to convert tyrosine into melanin has been compromised. Second, the enzyme might be incapable of any function, in which case the mutational event in the DNA of the gene would have produced an albinism allele, referred to earlier as an *a* allele. Hence a person of genotype *a*/*a* is an albino. The genotype *A*/*a* is interesting. It results in normal pigmentation because transcription of one copy of the wild-type allele (*A*) can provide enough tyrosinase for synthesis of normal amounts of melanin. Alleles are termed *haplosufficient* if roughly normal function is obtained when there is only a single copy of the normal gene. Alleles commonly appear to be haplosufficient, in part because small reductions in function are not vital to the organism. Alleles that fail to code for a functional protein are called null ("nothing") alleles and are generally not expressed in combination with functional alleles (in individuals of genotype *A*/*a*). The molecular basis of albinism is represented in figure.

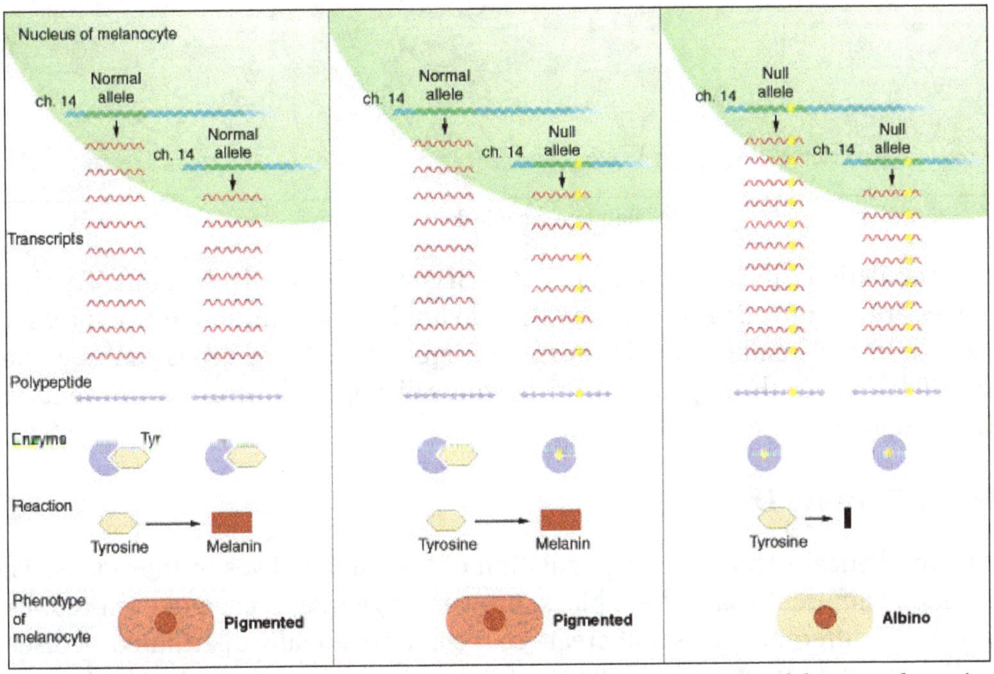

Molecular basis of albinism. Expression in cells containing 2, 1, and 0 copies of the normal tyrosinase allele on chromosome 14. Melanocytes are specialized melanin-producing cells.

The mutational site in the DNA can be of a number of types. The simplest and most common type is nucleotide-pair substitution, which can lead to amino acid substitution or to premature stop codons. Small deletions and duplication also are common. Even a single base deletion or insertion produces widespread damage at the protein level; because mRNA is read from one end "in frame" in groups of three, a loss or gain of one nucleotide pair shifts the reading frame, and all the amino acids translationally downstream will be incorrect. Such mutations are called frameshift mutations.

At the protein level, mutation changes the amino acid composition of the protein. The most important outcomes are change in shape and size. Such change in shape or size can result in no biological function (which would be the basis of a null allele), or reduced function. More rarely, mutation can lead to new function of the protein product.

Genetic Drift

Although variations of genes (also known as alleles) can be selected for because they help or hinder an organism, other mutations can have no effect. When the allele itself is not responsible for the change in its frequency in a population, genetic drift is acting on the allele. The graph below shows different trajectories for the same genes over time. As you can see, the frequency of these genes can change drastically over time, especially with the smallest populations.

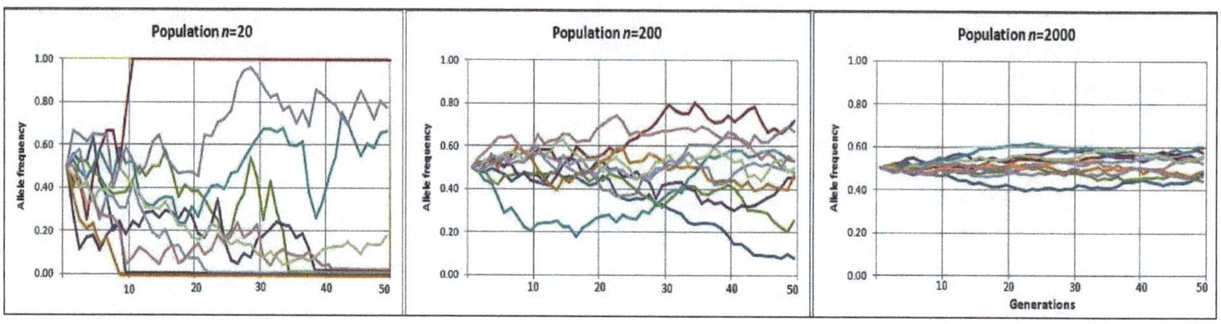

Random genetic drift chart.

In the largest populations, the allele frequency of each gene stays relatively stable. This happens because the genes are not affecting fitness, and thus do not have a natural selection pressure against or for the allele. In the smallest populations, the frequency of these genes can fluctuate greatly. Some become fixed within the population, while others disappear. These chance events which lead to changes in frequency are called genetic drift.

Examples of Genetic Drift

- In a hypothetical population: A population of 100 rabbits lives in the woods. The rabbits have many different coat colors: black, brown, tan, white, grey, and even red. In the population, the different alleles that create coat color are equally distributed. A disease comes into the rabbit population and kills 98 of the rabbits. The only rabbits that are left are red and grey rabbits, simply by chance. The genes have thus "drifted" from 6 alleles to only 2. This is an example of a bottleneck effect.
- In real life: Genetic drift happens all the time in populations, although it is not easily seen. Often, mutations arise that have little effect on the organism. These mutations get passed on if the organism reproduces, and do not get passed on if the organism does not survive. Although genetic drift used to be thought of in only small populations, even large populations experience genetic drift of certain alleles. This happens because a small number of individuals carry the alleles. Whether or not these alleles are duplicated is not a function of natural selection, but of chance. Many alleles come or go in populations without affecting great change.

Causes of Genetic Drift

Genetic drift is much more likely in smaller populations of organisms. The individual lines in the graph track the frequency of alleles in a given population. When the population is small and many alleles exist, any of the alleles can quickly become fixed or extinct in the population. When there are many organisms in the population, there is less of a chance of losing an entire allele, because many organisms carry the allele and it is less likely they will all be wiped out.

Genetic drift can easily be confused with natural selection. The difference is whether or not the allele is actively participating in the change in allele frequencies. If the allele affects an organism in a way that causes more reproduction of the DNA, the allele will increase in frequency. If it causes harm, it will decrease. This is caused by the allele's direct effects on the organism and the environment. This is natural selection. When the allele is increased or decreased simply because it was present in the random organisms that survived, this is genetic drift.

Types of Genetic Drift

1. Population Bottleneck: A population bottleneck is a type of genetic drift in which a population's size severely decreases. Competition, disease, or predation leads to these massive decreases in population size. The allele pool is now determined by the organisms which did not die. Some alleles increase in frequency simply because they are the only alleles left. This type of genetic drift can be seen when people don't take their entire course of antibiotics.

Antibiotics kill harmful bacteria in your system, regardless of what alleles they have. Antibiotics cause a massive reduction in harmful bacteria. This stops symptoms of the disease. A small population will survive if a patient quits their antibiotic early. This much smaller population could have allele frequencies that are very different from the original population of bacteria. These changes do not reflect the success or failure of the different alleles, but rather the effects of a random selection of bacteria. The new alleles will dominate the population until selection or more genetic drift cause the allele frequencies to change.

2. Founder Effect: In another type of genetic drift known as the founder effect, a new population is formed, or "founded", in a new location. If this new population does not interact and reproduce with the main population, the allele frequencies in this population will be much different from that of the parent population. Many islands contain species that only exist on a single island because of the founder effect.

For instance, if only two birds of a species land on an island, their alleles alone will account for the diversity present. While these alleles will dominate at first, mutations will arise in the population that will lead to new adaptations. This new adaptation stays with the founding population. With enough time, the two populations can diverge to a point which they can no longer interbreed. Species often separate in this way.

Genetic Drift Vs. Gene Flow

The concept of genetic drift is often confused with the concept of gene flow in biology. Gene flow is the movement of genes between populations, species, or between organisms. For instance, bacterial cells are able to transfer genes between different cells as a method of gaining antibiotic resistance. Populations of organisms exhibit gene flow when individuals from one population migrate and breed with a new population. Gene flow does not analyze the allele frequency of genes. Rather, it is a concept which describes the movement of genes between populations. By contrast, genetic drift describes the random selection of genes within a population, not attributable to natural selection forces.

References

- Laws-of-inheritance, boundless-biology: lumenlearning.com, Retrieved 05, April 2020
- Pleiotropy-definition-4687155: thoughtco.com, Retrieved 29, January 2020
- Haploid: biologydictionary.net, Retrieved 10, June 2020
- Euploidy-meaning-and-types-cell-biology, chromosomes-38885: biologydiscussion.com, Retrieved 06, March 2020
- Natural-selection, science: britannica.com, Retrieved 12, August 2020
- Genetic-drift: biologydictionary.net, Retrieved 25, February 2020

3

Gene Expression and Regulation

The process of formation of a functional gene product by using the information from a gene is termed as gene expression. Gene expression refers to the mechanisms that are used by the cells to regulate the production of specific gene products. This chapter has been carefully written to provide an easy understanding of the varied facets of gene expression and regulation.

Gene Expression

Gene expression is a process by which the genetic information is used in the synthesis of functional products including proteins and functional RNAs (e.g., tRNA, small nuclear RNA, microRNA, small/short interfering RNA, etc.) the process of gene expression is applied by all organisms including eukaryotes, prokaryotes, and viruses to produce the macromolecular machinery for life. Through controlling the cell structure and function, the gene plays an important role in cellular differentiation, morphogenesis, adaptability, and diversity. Because the control of the timing, location, and levels of gene expression can have a significant effect on gene functions in a single cell or a multicellular organism, gene regulation may also drive evolutionary change. Several steps in the gene expression process can be regulated, such as transcription, posttranscriptional modification (e.g., RNA splicing, 3′ poly A adding, 5′-capping), translation, and posttranslational modification (e.g., protein splicing, folding, and processing).

Transcription

The genomic DNA is composed of two antiparallel strands with 5′ and 3′ ends which are reverse and complementary for each. Regarding to a gene, the two DNA strands are classified as the "coding strand (sense strand)," which includes the DNA version of the RNA transcript sequence, and the "template strand (antisense strand, noncoding strand)" which serves as a blueprint for synthesizing an RNA strand. During transcription, the DNA template strand is read by an RNA polymerase to produce a complementary and antiparallel RNA primary transcript. Transcription is the first step of gene expression which involves copying a DNA sequence to make an RNA molecule including messenger RNA (mRNA), ribosome RNA (rRNA), and transfer RNA (tRNA) by the principle of complementary base pairing. In prokaryotes, transcription is performed by enzymes called RNA polymerases to form all RNA molecules. In eukaryotes, it is mainly performed by RNA polymerases I, II, III, IV, and V to make rRNA, mRNA, tRNA, etc.

Table: RNA polymerase function in eukaryotes.

RNA polymerase	Function
RNA polymerase I	Synthesis of precursor 45S rRNA (35S in yeast) which matures into 28S, 18S, and 5.8S rRNAs; the formation of the major RNA sections of the ribosome.

RNA polymerase II	Synthesis of mRNA precursors; most snRNA and miRNA.
RNA polymerase III	Synthesis of tRNAs, 5S rRNA and other small RNAs found in the nucleus and cytosol.
RNA polymerase IV	Synthesis of siRNA in plants.
RNA polymerase V	Synthesis of RNAs involved in siRNA-directed heterochromatin formation in plants.
Abbreviation: mRNA, messenger RNA; rRNA, ribosome RNA; tRNA, transfer RNA; snRNA, small nuclear RNA; miRNA, microRNA; siRNA, small/short interfering RNA.	

Post-transcriptional Modification

Post-transcriptional modification is referred to as biological processes by which RNA primary transcripts are chemically changed posterior to transcription in eukaryotes. This process significantly modifies the chemical structure of RNA molecules by three main constituting steps: the putting of a 5′ cap, the addition of a 3′ polyadenylated tail, and RNA splicing. In eukaryotes, primary RNA transcript must be processed after transcription because the initial precursor mRNA (primary RNA transcript) usually contains both exons (coding sequences) and introns (noncoding sequences). By removing the introns to link the exons directly in RNA splicing, such processing is able to ensure the correct translation of eukaryotic genomes. Also, posttranscription modification can protect primary RNA transcripts from degradation in that the 5′ cap and poly-A tail are able to protect the transcripts and facilitate the transportation of the mRNAs to ribosomes.

Translation

Every mRNA is composed of a 5′ untranslated region (5′UTR), a protein-coding region or open reading frame (ORF), and a 3′ untranslated region (3′UTR). The ORF carries information for protein synthesis encoded by the RNA triplet codes. The codon, a RNA triplet of the coding region, corresponds to a binding site complementary to an anticodon triplet in tRNAs. tRNAs with different anticodon triplets carry different amino acids and thereby make amino acids link together according to the order of triplets in the coding region by the assistance of ribosomes.

In prokaryotes, translation usually occurs in the cytosol where the specific amino acid, enzymes, and small subunits of the ribosome bind to the tRNA. Translation simultaneously proceeds with transcription (co-transcriptionally), using a mRNA that is still being synthesized. In eukaryotes, transcription occurs in nucleus, and translation occurs in cytoplasm (not co-transcriptionally). Although translation can occur in a variety of regions of the eukaryotic cells, its major working locations are the cytoplasm for soluble cytoplasmic proteins and the endoplasmic reticulum (ER) membrane for secretory proteins or integral proteins.

Post-translational modification is referred as the chemical change which proteins may undergo after translation. The main alternations are the specific cleavage of precursor proteins, the excision of signal recognition peptides (SRP), the formation of disulfide bonds, the covalent addition or removal of low-molecular-weight groups, and the addition of metal ions, leading to protein modifications such as glycosylation, lipidation, hydroxylation, methylation, mono-ADP-ribosylation, myristoylation, oxidation, palmitoylation, and phosphorylation. It plays a crucial role in the regulation of the protein folding, the targeting of specific subcellular compartments, the interaction of ligands or other proteins, and the state of functions, such as enzyme catalytic activity or the signaling function of proteins in signal transduction pathways.

Reverse Transcription

Reverse transcription is referred to as the synthesis of complementary DNA (cDNA) from an RNA template using reverse transcriptase (RTase). For reverse transcription, RT uses an RNA template and a short RNA primer complementary to the 3′ end of the RNA to direct a cDNA strand synthesis from 5′ end to 3′ end. The first-strand cDNA can be made to be double-stranded using DNA polymerase I, and ribonuclease (RNase) H activity for RNA primer digestion is required in this case. The engineered RTase can improve the efficiency of full-length cDNA products, ensure the completeness of the mRNA transcript copying, and enable the propagation of a faithful DNA copy of an RNA sequence. The use of the thermostable RTase is very helpful for the denaturation of RNA structure when the RNA sequence contains high amounts of secondary structure. Some retroviruses (e.g., human immunodeficiency virus, HIV) which have an RNA genome are able to transcribe RNA into cDNA using RTase. The RNase H digests the RNA strand, and then the cDNA strand is used as a template to synthesize a complementary DNA strand to form a double helix DNA structure. The resulting double stranded DNA (ds DNA) can be integrated into the host DNA genome, causing the host cell to produce viral proteins that assemble into new virions. In HIV, the host cell undergoes apoptosis (programmed cell death).

Telomerase (terminal transferase, RNA-directed DNA polymerase) is a kind of RTase that lengthens the ends of linear chromosomes in some eukaryotes, containing an RNA template from which it synthesizes a repeating sequence of DNA or "junk" DNA. It is active in stem cells, gametes. and most cancer cells to enable these cells to replicate their genomes immortally without losing important protein-coding DNA sequence. Embryonic stem cells highly express telomerase to allow these cells to divide repeatedly and form the individual such as male sperm cells, epidermal cells, and activated T and B lymphocytes. However, it is usually absent from or at very low levels in most somatic cells. The repeated DNA sequence (telomere) can be considered as a "cap" for a chromosome. In normal cells, the telomere is shortened when a linear chromosome is duplicated. Activation of telomerase is one of the processes that let cancer cells become indefinite. Telomerase allows each offspring to avoid losing a bit of DNA, making the normal cells divide without limitation and become abnormal cells, and the unbounded cell growth is a characteristic of cancer.

Epigenetics

Epigenetics is the study of heritable phenotype changes that do not involve alterations in the DNA sequence, meaning a change in phenotype without a change in genotype. The term also refers to the functionally associated changes to the genome that do not encompass a change in the nucleotide sequence. At present, DNA methylation, histone modification, and noncoding RNA (ncRNA)-associated gene silencing are major functions for involving in the initiation and support of epigenetic changes. It often indicates changes that affect gene activity and expression and shows phenotypic changes which can be transferred to the offspring.

Epigenetic changes can be influenced by several factors including the age, environment, lifestyle, and disease; it is traditionally considered to be regular and natural. This kind of change may be consistent through cell divisions during the cell life cycle and may also sustain for several generations, even if they do not have changes in the organism's underlying DNA sequence. The process of cellular differentiation is an example of an epigenetic change in eukaryotic biology.

The Evolving Landscape of Epigenetic Research

In 1969, Griffith and Mahler first suggested that DNA methylation might play a crucial role in long-term memory function. DNA methylation has currently become one of the most extensively studied and well-characterized epigenetic modifications. The other main modifications include chromatin remodeling, histone modifications, and noncoding RNA mechanisms. The new findings about epigenetics are the correlation between epigenetic changes and diseases such as cancers, mental retardation, immune disorders, neuropsychiatric disorders, and pediatric disorders.

Environment and Lifestyle can Influence Epigenetic Change from One Generation to the Next

Not only the environment but also individual lifestyles can directly interact with the genome to affect epigenetic changes. These epigenetic changes may be demonstrated at various stages throughout an individual's life and even last to his offspring. The prenatal and early postnatal environmental factors can influence the adult risk for the incidence of various chronic diseases and behavioral disorders in human epidemiology studies. It is known that the children have elevated rates of coronary heart disease and obesity after their mothers are exposed to famine during early pregnancy compared with those who are not exposed. Additionally, maternal exposure to air pollution could affect her children's asthma susceptibility. Similarly, the adults who were prenatally exposed to famine have also been reported to have significantly higher incidence of schizophrenia. Fortunately, maternal ingestion of vitamin D is capable to adjust DNA methylation that impacts placenta function.

Environment and Lifestyle affect Individual Epigenetics and Health

Epigenetics are considered to be dynamic and changeable by the influence of lifestyle options and environmental factors, though our epigenetic marks are more stable during adulthood. Epigenetic effects gradually occur both in the womb and the full course of a human life span, and epigenetic changes could be reversed. Epigenetics have shown that different lifestyle options and environmental exposures can change DNA marks and play a vital role in the determination of health outcomes. The environment can dominantly influence the epigenetic tags and disease susceptibility. Pollution has become a significant topic because scientists have found air pollution could induce DNA methylation and increase one's risk for neurodegenerative disease. Fortunately, vitamin B groups potentially protect humans from harmful epigenetic effects of pollution and against the other harmful effects on the body.

Deciphering the Relationship between Epigenetics and Diseases

It is known that chronic pancreatitis causes a high risk of inflammation-associated progression to pancreatic cancer. The difficulty in rapidly diagnosing the disease is closely associated with its high mortality rate. Previous studies have demonstrated that cell-free DNA methylation from inflammatory diseases or cancer is variable, thereby opening a new era in developing biomarkers for the early diagnosis of diseases. Hence, early diagnosis for pancreatic cancer becomes crucial and facilitates the related studies into the epigenetic profiles. Researcher reported that exploiting the relationship between abnormally methylated cell-free DNA and pre-neoplastic lesions or chronic pancreatitis may become a novel method in developing tools for the early diagnosis of pancreatic cancer. Early diagnosis potentially makes it possible for the prediction of prognosis, the monitoring of tumor progression, and the development of effectively therapeutic strategies and provides precision medicine for patients suffering from a pancreatic disease.

The early-life environment including air quality is known to be critical for fetal programming. The air pollution exposure to mothers during pregnancy may adversely influence new born outcomes such as baby birth weight, preternatural birth, and preterm birth. Therefore, it is needed to understand both air pollution-induced early health effects and its later-life consequences. Researcher provided an overview of air pollution-induced placental molecular changes observed in the Environage birth cohort and assess the existing evidence. They reported that nitrosative stress and epigenetic alterations in the placenta may result from the prenatal exposure to air pollution. It is crucial to realize the clinical consequences of early-life epigenetic changes in the follow-up of child or birth cohort study. The public health policy maker should have understanding of epigenetic consequences and transgenerational risks to propose effective strategies which are focusing on providing effective protection of pregnant women, unborn children, and infants against exposure to adverse lifestyle factors.

Phenotypic Traits

Genotype, a unique genome that can be revealed by genomic sequencing is a complete inheritable genetic identity. It is mediated by a special gene, cluster of genes, or set of genes which are carried by an individual. Genes are certain DNA segments that code for the protein production to determine distinct traits of individuals. DNA contains the genetic code which is responsible for all cellular functions such as mitosis, meiosis, DNA replication, protein synthesis, molecule transportation, etc. Each gene is located on a chromosome and can exist in different forms called alleles which are located on specific regions of chromosomes. The alleles can be transmitted from parents to offspring through sexual reproduction. The diploid organism inherits two alleles for each gene; one allele is from the father, and the other allele is from the mother. The interactions between alleles determine an organism's phenotype.

The phenotype is a description of actual physical characteristics and encompasses directly visible characteristics such as height, weight, skin color, eye color, size, shape, health condition, disease history, even behavior, and temperament. However, not all phenotypes are a direct result of genotype. Most phenotypes are affected by both the genotype and environment in which one has lived one's life including everything that has ever happened. We often consider "nature" as the unique genome which one carries and "nurture" as the environment in which one has lived one's life. The phenotype is dependent on the genetic makeup of the organism and influenced by the environment to which the organism is subjected across various epigenetic processes. It includes all of the organism's characteristics, including traits at multiple levels of biological organization, ranging from individual behavior and trait evolution through morphology, physiology, cellular characteristics, biochemical pathways, and gene expression.

A phenotypic trait (simply trait or character state) is a distinct variant of an organism's characteristic, and it is an obvious and measurable trait that is expressed in an observable way. For example, the eye color (green, blue, brown, and hazel) is a phenotypic trait which is a polygenetic inheritance. The phenotypic trait may be either inherited parentally or determined environmentally; that is to say, some traits are determined by the genotype, and some traits are determined by environmental factors. The different genes or alleles caused by mutation can be passed on to successive generations, resulting in different phenotypic traits. Though the environment can affect the phenotype, the heritability of a phenotypic trait is defined as the proportion of the total phenotypic variation of this specific trait that is elucidated only by the genetic variation. The phenotypic variation of a trait (P) can be divided into three contributions as follows: latent genetic (G) factors, environmental (E) factors, and gene–environment interactions (G × E).

If an organism inherits two same alleles, it is homozygous and expresses only one phenotypic trait. If an organism inherits two different alleles, it is heterozygous and may express more than one phenotypic trait. The phenotypic traits can be dominant or recessive. In completely dominant inheritance, only dominant traits are observable because the phenotype of the dominant trait entirely masks the phenotypc of the recessive trait. In contrast, the dominant allele does not mask the other allele completely in incomplete dominance inheritance, resulting in a phenotype that is a mixture of traits from both alleles. In codominance relationships, both alleles are fully expressed, resulting in a phenotype in which both traits are demonstrated independently.

Gene Regulation

Gene regulation is the process of turning genes on and off. During early development, cells begin to take on specific functions. Gene regulation ensures that the appropriate genes are expressed at the proper times. Gene regulation can also help an organism respond to its environment. Gene regulation is accomplished by a variety of mechanisms including chemically modifying genes and using regulatory proteins to turn genes on or off.

Each nucleated cell in a multicellular organism contains copies of the same DNA. Similarly, all cells in two pure bacterial cultures inoculated from the same starting colony contain the same DNA, with the exception of changes that arise from spontaneous mutations. If each cell in a multicellular organism has the same DNA, then how is it that cells in different parts of the organism's body exhibit different characteristics? Similarly, how is it that the same bacterial cells within two pure cultures exposed to different environmental conditions can exhibit different phenotypes? In both cases, each genetically identical cell does not turn on, or express, the same set of genes. Only a subset of proteins in a cell at a given time is expressed.

Genomic DNA contains both structural genes, which encode products that serve as cellular structures or enzymes, and regulatory genes, which encode products that regulate gene expression. The expression of a gene is a highly regulated process. Whereas regulating gene expression in multicellular organisms allows for cellular differentiation, in single-celled organisms like prokaryotes, it primarily ensures that a cell's resources are not wasted making proteins that the cell does not need at that time.

Elucidating the mechanisms controlling gene expression is important to the understanding of human health. Malfunctions in this process in humans lead to the development of cancer and other diseases. Understanding the interaction between the gene expression of a pathogen and that of its human host is important for the understanding of a particular infectious disease. Gene regulation involves a complex web of interactions within a given cell among signals from the cell's environment, signaling molecules within the cell, and the cell's DNA. These interactions lead to the expression of some genes and the suppression of others, depending on circumstances.

Prokaryotes and eukaryotes share some similarities in their mechanisms to regulate gene expression; however, gene expression in eukaryotes is more complicated because of the temporal and spatial separation between the processes of transcription and translation. Thus, although most regulation of gene expression occurs through transcriptional control in prokaryotes, regulation of gene expression in eukaryotes occurs at the transcriptional level and post-transcriptionally (after the primary transcript has been made).

Prokaryotic Gene Regulation

In bacteria and archaea, structural proteins with related functions are usually encoded together within the genome in a block called an operon and are transcribed together under the control of a single promoter, resulting in the formation of a polycistronic transcript. In this way, regulation of the transcription of all of the structural genes encoding the enzymes that catalyze the many steps in a single biochemical pathway can be controlled simultaneously, because they will either all be needed at the same time, or none will be needed. For example, in *E. coli*, all of the structural genes that encode enzymes needed to use lactose as an energy source lie next to each other in the lactose (or *lac*) operon under the control of a single promoter, the *lac* promoter. French scientists François Jacob and Jacques Monod at the Pasteur Institute were the first to show the organization of bacterial genes into operons, through their studies on the *lac* operon of *E. coli*. For this work, they won the Nobel Prize in Physiology or Medicine in 1965. Although eukaryotic genes are not organized into operons, prokaryotic operons are excellent models for learning about gene regulation generally. There are some gene clusters in eukaryotes that function similar to operons. Many of the principles can be applied to eukaryotic systems and contribute to our understanding of changes in gene expression in eukaryotes that can result pathological changes such as cancer.

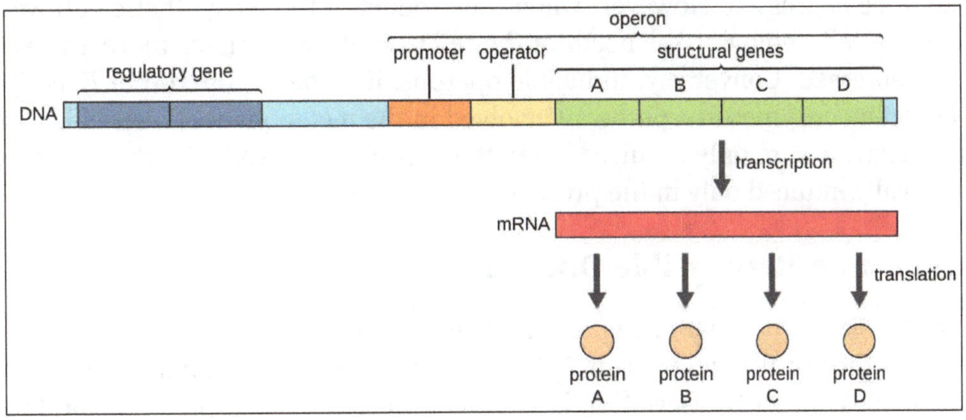

In prokaryotes, structural genes of related function are often organized together on the genome and transcribed together under the control of a single promoter. The operon's regulatory region includes both the promoter and the operator. If a repressor binds to the operator, then the structural genes will not be transcribed. Alternatively, activators may bind to the regulatory region, enhancing transcription.

Each operon includes DNA sequences that influence its own transcription; these are located in a region called the regulatory region. The regulatory region includes the promoter and the region surrounding the promoter, to which transcription factors, proteins encoded by regulatory genes, can bind. Transcription factors influence the binding of RNA polymerase to the promoter and allow its progression to transcribe structural genes. A repressor is a transcription factor that suppresses transcription of a gene in response to an external stimulus by binding to a DNA sequence within the regulatory region called the operator, which is located between the RNA polymerase binding site of the promoter and the transcriptional start site of the first structural gene. Repressor binding physically blocks RNA polymerase from transcribing structural genes. Conversely, an activator is a transcription factor that increases the transcription of a gene in response to an external stimulus by facilitating RNA polymerase binding to the promoter. An inducer, a third type of regulatory molecule, is a small molecule that either activates or represses transcription by interacting with a repressor or an activator.

In prokaryotes, there are examples of operons whose gene products are required rather consistently and whose expression, therefore, is unregulated. Such operons are constitutively expressed, meaning they are transcribed and translated continuously to provide the cell with constant intermediate levels of the protein products. Such genes encode enzymes involved in housekeeping functions required for cellular maintenance, including DNA replication, repair, and expression, as well as enzymes involved in core metabolism. In contrast, there are other prokaryotic operons that are expressed only when needed and are regulated by repressors, activators, and inducers.

Regulation by Repression

Prokaryotic operons are commonly controlled by the binding of repressors to operator regions, thereby preventing the transcription of the structural genes. Such operons are classified as either repressible operons or inducible operons. Repressible operons, like the tryptophan (*trp*) operon, typically contain genes encoding enzymes required for a biosynthetic pathway. As long as the product of the pathway, like tryptophan, continues to be required by the cell, a repressible operon will continue to be expressed. However, when the product of the biosynthetic pathway begins to accumulate in the cell, removing the need for the cell to continue to make more, the expression of the operon is repressed. Conversely, inducible operons, like the *lac* operon of *E. coli*, often contain genes encoding enzymes in a pathway involved in the metabolism of a specific substrate like lactose. These enzymes are only required when that substrate is available, thus expression of the operons is typically induced only in the presence of the substrate.

The *trp* Operon: A Repressible Operon

E. coli can synthesize tryptophan using enzymes that are encoded by five structural genes located next to each other in the *trp* operon. When environmental tryptophan is low, the operon is turned on. This means that transcription is initiated, the genes are expressed, and tryptophan is synthesized. However, if tryptophan is present in the environment, the *trp* operon is turned off. Transcription does not occur and tryptophan is not synthesized.

When tryptophan is not present in the cell, the repressor by itself does not bind to the operator; therefore, the operon is active and tryptophan is synthesized. However, when tryptophan accumulates in the cell, two tryptophan molecules bind to the *trp* repressor molecule, which changes its shape, allowing it to bind to the *trp* operator. This binding of the active form of the *trp* repressor to the operator blocks RNA polymerase from transcribing the structural genes, stopping expression of the operon. Thus, the actual product of the biosynthetic pathway controlled by the operon regulates the expression of the operon.

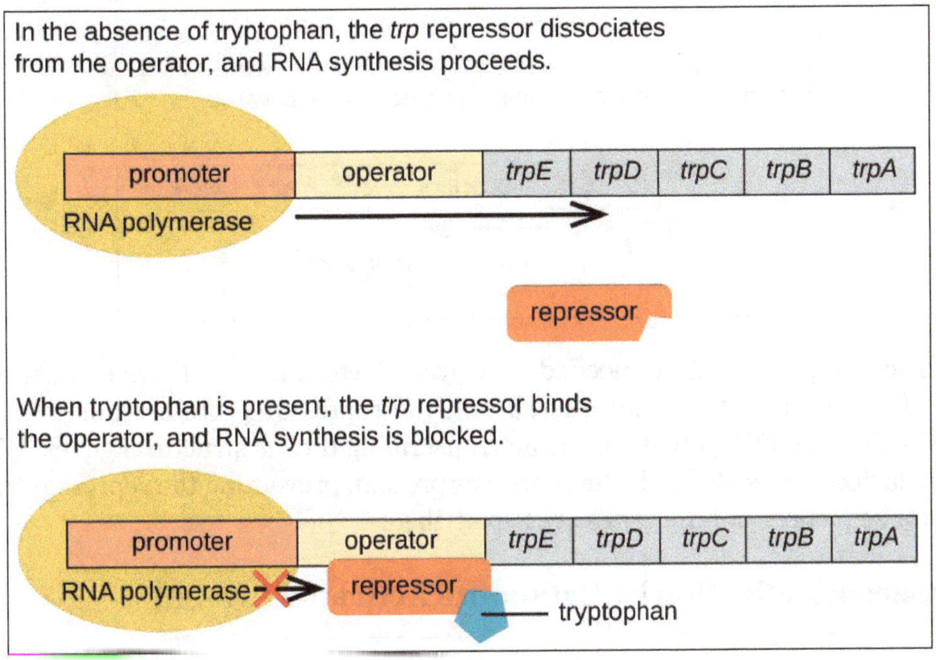

The five structural genes needed to synthesize tryptophan in *E. coli* are located next to each other in the *trp* operon. When tryptophan is absent, the repressor protein does not bind to the operator, and the genes are transcribed. When tryptophan is plentiful, tryptophan binds the repressor protein at the operator sequence. This physically blocks the RNA polymerase from transcribing the tryptophan biosynthesis genes.

The *lac* Operon: An Inducible Operon

The *lac* operon is an example of an inducible operon that is also subject to activation in the absence of glucose. The *lac* operon encodes three structural genes necessary to acquire and process the disaccharide lactose from the environment, breaking it down into the simple sugars glucose and galactose. For the *lac* operon to be expressed, lactose must be present. This makes sense for the cell because it would be energetically wasteful to create the enzymes to process lactose if lactose was not available.

In the absence of lactose, the *lac* repressor is bound to the operator region of the *lac* operon, physically preventing RNA polymerase from transcribing the structural genes. However, when lactose is present, the lactose inside the cell is converted to allolactose. Allolactose serves as an inducer molecule, binding to the repressor and changing its shape so that it is no longer able to bind to the operator DNA. Removal of the repressor in the presence of lactose allows RNA polymerase to move through the operator region and begin transcription of the *lac* structural genes.

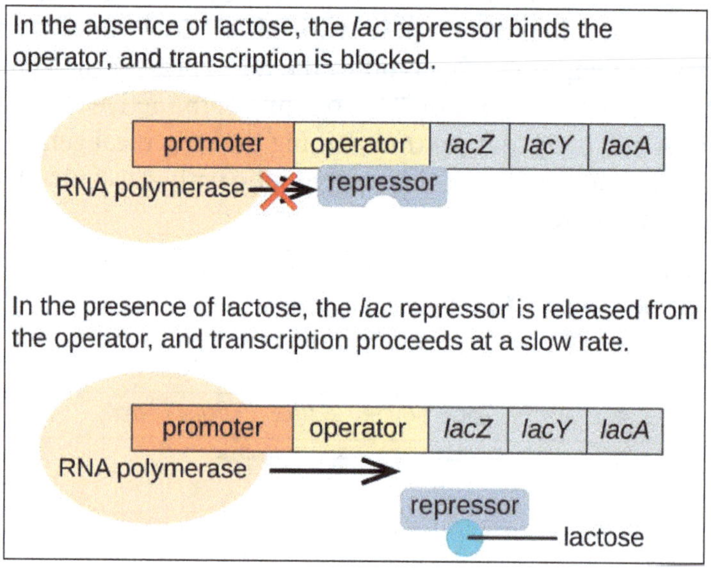

The three structural genes that are needed to degrade lactose in *E. coli* are located next to each other in the *lac* operon. When lactose is absent, the repressor protein binds to the operator, physically blocking the RNA polymerase from transcribing the *lac* structural genes. When lactose is available, a lactose molecule binds the repressor protein, preventing the repressor from binding to the operator sequence, and the genes are transcribed.

The *lac* Operon: Activation by Catabolite Activator Protein

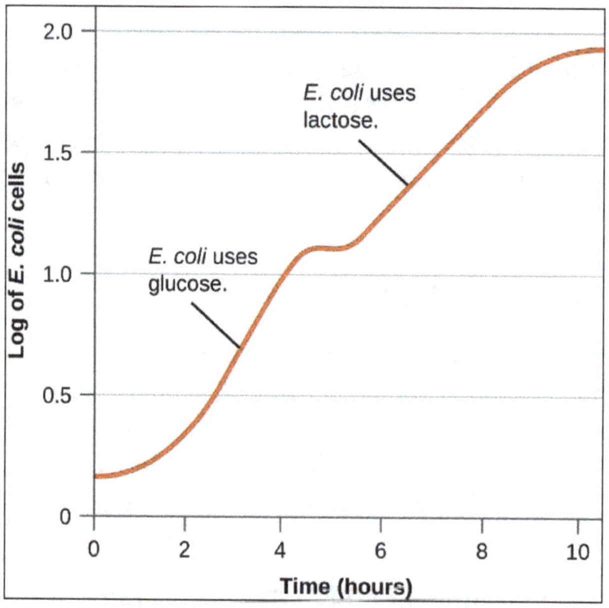

When grown in the presence of two substrates, *E. coli* uses the preferred substrate (in this case glucose) until it is depleted. Then, enzymes needed for the metabolism of the second substrate are expressed and growth resumes, although at a slower rate.

Bacteria typically have the ability to use a variety of substrates as carbon sources. However, because glucose is usually preferable to other substrates, bacteria have mechanisms to ensure that alternative substrates are only used when glucose has been depleted. Additionally, bacteria have mechanisms to ensure that the genes encoding enzymes for using alternative substrates are expressed only when the alternative substrate is available. In the 1940s, Jacques Monod was the first to demonstrate the preference for certain substrates over others through his studies of *E. coli*'s growth when cultured in the presence of two different substrates simultaneously. Such studies generated diauxic growth curves, like the one shown in figure. Although the preferred substrate glucose is used first, *E. coli* grows quickly and the enzymes for lactose metabolism are absent. However, once glucose levels are depleted, growth rates slow, inducing the expression of the enzymes needed for the metabolism of the second substrate, lactose. Notice how the growth rate in lactose is slower, as indicated by the lower steepness of the growth curve.

The ability to switch from glucose use to another substrate like lactose is a consequence of the activity of an enzyme called Enzyme IIA (EIIA). When glucose levels drop, cells produce less ATP from catabolism, and EIIA becomes phosphorylated. Phosphorylated EIIA activates adenylyl cyclase, an enzyme that converts some of the remaining ATP to cyclic AMP (cAMP), a cyclic derivative of AMP and important signaling molecule involved in glucose and energy metabolism in *E. coli*. As a result, cAMP levels begin to rise in the cell.

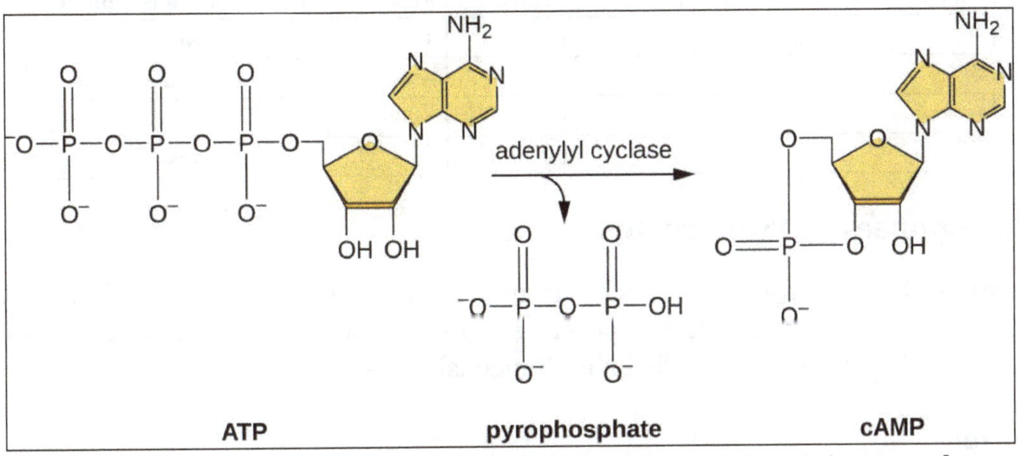

When ATP levels decrease due to depletion of glucose, some remaining ATP is converted to cAMP by adenylyl cyclase. Thus, increased cAMP levels signal glucose depletion.

The *lac* operon also plays a role in this switch from using glucose to using lactose. When glucose is scarce, the accumulating cAMP caused by increased adenylyl cyclase activity binds to catabolite activator protein (CAP), also known as cAMP receptor protein (CRP). The complex binds to the promoter region of the *lac* operon. In the regulatory regions of these operons, a CAP binding site is located upstream of the RNA polymerase binding site in the promoter. Binding of the CAP-cAMP complex to this site increases the binding ability of RNA polymerase to the promoter region to initiate the transcription of the structural genes. Thus, in the case of the *lac* operon, for transcription to occur, lactose must be present (removing the lac repressor protein) and glucose levels must be depleted (allowing binding of an activating protein). When glucose levels are high, there is catabolite repression of operons encoding enzymes for the metabolism of alternative substrates. Because of low cAMP levels under these conditions, there is an insufficient amount of the CAP-cAMP complex to activate transcription of these operons.

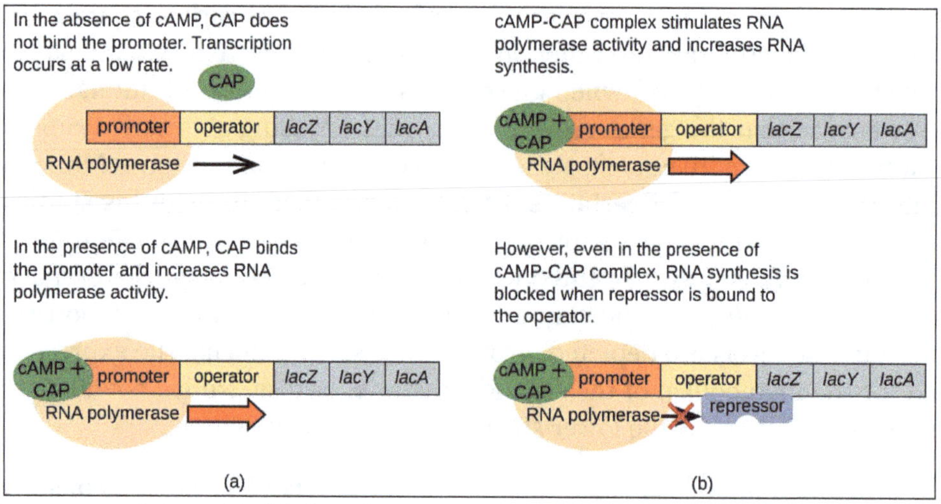

(a) In the presence of cAMP, CAP binds to the promoters of operons, like the lac operon, that encode genes for enzymes for the use of alternate substrates. (b) For the lac operon to be expressed, there must be activation by cAMP-CAP as well as removal of the lac repressor from the operator.

Table: Conditions affecting transcription of the *lac* Operon.

Glucose	CAP binds	Lactose	Repressor binds	Transcription
+	–	–	+	No
+	–	+	–	Some
–	+	–	+	No
–	+	+	–	Yes

Global Responses of Prokaryotes

In prokaryotes, there are also several higher levels of gene regulation that have the ability to control the transcription of many related operons simultaneously in response to an environmental signal. A group of operons all controlled simultaneously is called a regulon.

Alarmones

When sensing impending stress, prokaryotes alter the expression of a wide variety of operons to respond in coordination. They do this through the production of alarmones, which are small intracellular nucleotide derivatives. Alarmones change which genes are expressed and stimulate the expression of specific stress-response genes. The use of alarmones to alter gene expression in response to stress appears to be important in pathogenic bacteria. On encountering host defense mechanisms and other harsh conditions during infection, many operons encoding virulence genes are upregulated in response to alarmone signaling. Knowledge of these responses is key to being able to fully understand the infection process of many pathogens and to the development of therapies to counter this process.

Alternate σ Factors

Since the σ subunit of bacterial RNA polymerase confers specificity as to which promoters should be transcribed, altering the σ factor used is another way for bacteria to quickly and globally change what regulons are transcribed at a given time. The σ factor recognizes sequences within a bacterial promoter, so different σ factors will each recognize slightly different promoter sequences. In this way, when the cell senses specific environmental conditions, it may respond by changing which σ factor it expresses, degrading the old one and producing a new one to transcribe the operons encoding genes whose products will be useful under the new environmental condition. For example, in sporulating bacteria of the genera *Bacillus* and *Clostridium* (which include many pathogens), a group of σ factors controls the expression of the many genes needed for sporulation in response to sporulation-stimulating signals.

Additional Methods of Regulation in Bacteria: Attenuation and Riboswitches

Although most gene expression is regulated at the level of transcription initiation in prokaryotes, there are also mechanisms to control both the completion of transcription as well as translation concurrently. Since their discovery, these mechanisms have been shown to control the completion of transcription and translation of many prokaryotic operons. Because these mechanisms link the regulation of transcription and translation directly, they are specific to prokaryotes, because these processes are physically separated in eukaryotes.

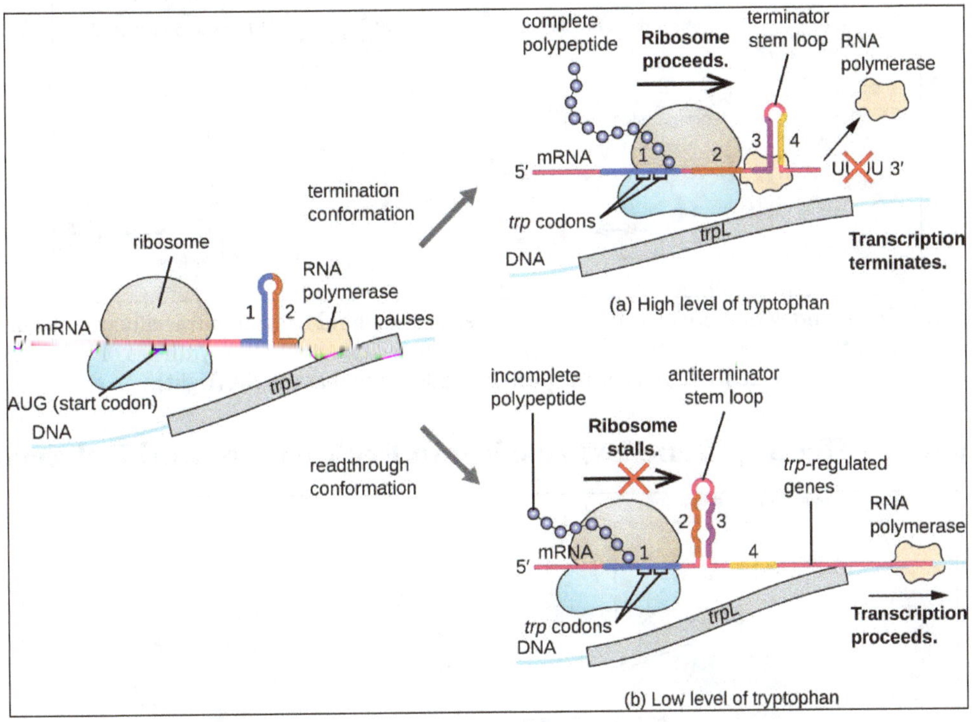

When tryptophan is plentiful, translation of the short leader peptide encoded by *trpL* proceeds, the terminator loop between regions 3 and 4 forms, and transcription terminates. When tryptophan levels are depleted, translation of the short leader peptide stalls at region 1, allowing regions 2 and 3 to form an antiterminator loop, and RNA polymerase can transcribe the structural genes of the *trp* operon.

One such regulatory system is attenuation, whereby secondary stem-loop structures formed within the 5' end of an mRNA being transcribed determines if transcription to complete the synthesis of this mRNA will occur and if this mRNA will be used for translation. Beyond the transcriptional repression mechanism, attenuation also controls expression of the *trp* operon in *E. coli*. The *trp* operon regulatory region contains a leader sequence called *trpL* between the operator and the first structural gene, which has four stretches of RNA that can base pair with each other in different combinations. When a terminator stem-loop forms, transcription terminates, releasing RNA polymerase from the mRNA. However, when an antiterminator stem-loop forms, this prevents the formation of the terminator stem-loop, so RNA polymerase can transcribe the structural genes.

A related mechanism of concurrent regulation of transcription and translation in prokaryotes is the use of a riboswitch, a small region of noncoding RNA found within the 5' end of some prokaryotic mRNA molecules. A riboswitch may bind to a small intracellular molecule to stabilize certain secondary structures of the mRNA molecule. The binding of the small molecule determines which stem-loop structure forms, thus influencing the completion of mRNA synthesis and protein synthesis.

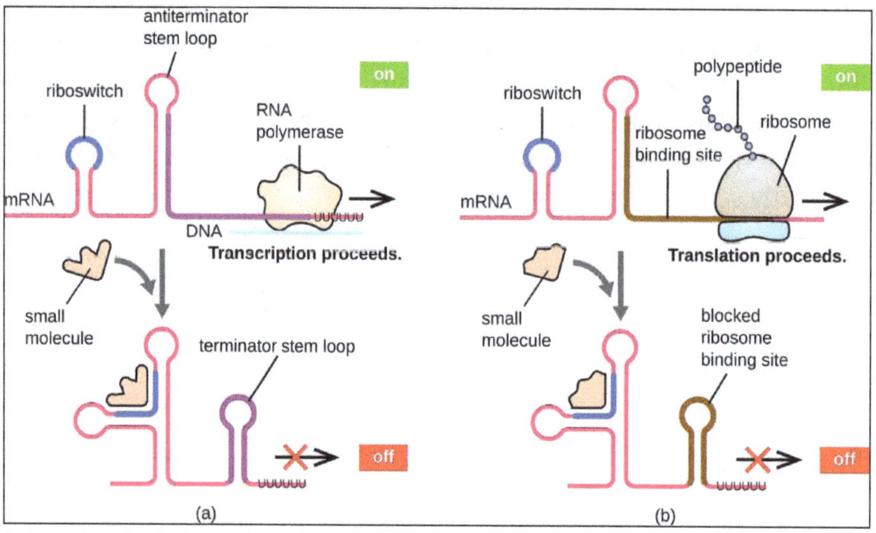

Riboswitches found within prokaryotic mRNA molecules can bind to small intracellular molecules, stabilizing certain RNA structures, influencing either the completion of the synthesis of the mRNA molecule itself (left) or the protein made using that mRNA (right).

Other Factors Affecting Gene Expression in Prokaryotes and Eukaryotes

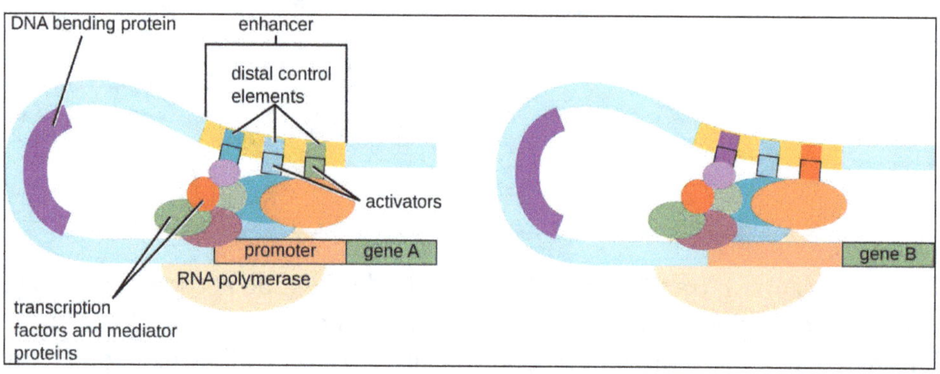

In eukaryotes, an enhancer is a DNA sequence that promotes transcription. Each enhancer is made up of short DNA sequences called distal control elements. Activators bound to the distal control elements interact with mediator proteins and transcription factors. Two different genes may have the same promoter but different distal control elements, enabling differential gene expression.

As in prokaryotes, eukaryotic transcription can be controlled through the binding of transcription factors including repressors and activators. Interestingly, eukaryotic transcription can be influenced by the binding of proteins to regions of DNA, called enhancers, rather far away from the gene, through DNA looping facilitated between the enhancer and the promoter. Overall, regulating transcription is a highly effective way to control gene expression in both prokaryotes and eukaryotes. However, the control of gene expression in eukaryotes in response to environmental and cellular stresses can be accomplished in additional ways without the binding of transcription factors to regulatory regions.

DNA Level Control

In eukaryotes, the DNA molecules or associated histones can be chemically modified in such a way as to influence transcription; this is called epigenetic regulation. Methylation of certain cytosine nucleotides in DNA in response to environmental factors has been shown to influence use of such DNA for transcription, with DNA methylation commonly correlating to lowered levels of gene expression. Additionally, in response to environmental factors, histone proteins for packaging DNA can also be chemically modified in multiple ways, including acetylation and deacetylation, influencing the packaging state of DNA and thus affecting the availability of loosely wound DNA for transcription. These chemical modifications can sometimes be maintained through multiple rounds of cell division, making at least some of these epigenetic changes heritable.

References

- Introductory-chapter-gene-expression-and-phenotypic-traits, gene-expression-and-phenotypic-traits: intechopen.com, Retrieved 29, April 2020
- Gene-regulation-operon-theory, microbiology: lumenlearning.com, Retrieved 10, February 2020

4

Understanding Genetic Engineering

Genetic engineering is a set of technologies that are used to manipulate the genes of an organism. A few of the techniques used in genetic engineering are cloning, transgenic technology, gene therapy, gene targeting, etc. The aim of this chapter is to explore these techniques of genetic engineering which are crucial for a complete understanding of the subject.

Genetic engineering is the process of modifying an organism's DNA to introduce new, desirable traits. Many organisms, from bacteria to plants and animals, have been genetically modified for academic, medical, agricultural, and industrial purposes. While genetic engineering has definite benefits, ethical concerns surround modifying humans and our food supply.

Scientists can Deliberately Modify an Organism's Genome

Genetic engineering is possible because the genetic code—the way information is encoded by DNA—and the structure of DNA are universal among all life forms. As a result, an organism's genetic code may be modified in several ways. The nucleotide sequence may be selectively edited by using techniques such as the CRISPR/Cas9 system. Known as the "molecular scissors," the CRISPR/Cas9 system is an innate, prokaryotic immune response that has been co-opted for editing genetic information.

A gene may also be removed from an organism to create a "knockout," or introduced to create a "knockin," through a process called gene targeting. This method relies on homologous recombination—genetic exchange between DNA molecules that share an extended region with similar sequences—to modify an endogenous gene. Scientists can also insert a gene from one organism into the genome of another, resulting in a transgenic organism. Generally, DNA combined from different sources is called recombinant DNA. The organism that receives that DNA is considered a genetically modified organism or GMO.

Genetic Engineering Impacts Human Health and Well-being

Genome editing has significantly impacted scientific research, agriculture, industry, and medicine. Molecular biology research often inserts transgenes—foreign genes—into bacteria and viruses to study gene function and expression. Bacteria were the first organisms to be genetically engineered. Scientists introduced the human insulin gene to produce synthetic insulin that is used by people with diabetes. A technique called gene therapy allows a new gene to be inserted into a person so that the protein it encodes can be expressed within their cells. Gene therapy provides a cure or treatment for some serious and otherwise untreatable genetic diseases. Scientists modified viruses to deliver new genes to host cells. These customized viruses can infect diseased cells and insert a correct copy of a defective gene, treating human disorders such as Severe Combined Immunodeficiency (SCID).

Although many gene therapy treatments use modified viruses, the CRISPR/Cas9 system has become an increasingly popular technique. The CRISPR/Cas9 system cuts DNA by using a guide—RNA sequences known as CRISPR—to direct the "molecular scissors"—an enzyme called Cas9—to specific sites in the genome. Scientists use this molecular tool to add, remove, or alter genetic material. CRISPR/Cas9 has been used in mouse models to correct errors in genes that are responsible for Duchenne muscular dystrophy, Hepatitis B, cataracts, and cardiovascular disease. While genetic engineering can yield new treatments for diseases, it can also be used for other practical purposes. Transgenic goats have been developed that produce spider silk in their milk for industrial use. In agriculture, some plants have been genetically modified to improve characteristics such as nutritional content and pest resistance. Recent and future advances in genetic engineering will likely continue to impact both human health and well-being.

Ethical Concerns Regarding Genetic Engineering

Genetic engineering has great potential, but where do we draw the line? Scientists and society must answer this question. Human genome editing, especially in germline cells, is a major ethical concern. Most gene therapies modify somatic cells, so genetic changes only affect the individual. Changes to a person's germline, however, are also inherited by their offspring.

Cloning

In general, cloning means the creation of a perfect replica. Typically, the word is used to describe the creation of a genetically identical copy. In biology, the re-creation of a whole organism is referred to as "reproductive cloning." Long before attempts were made to clone an entire organism, researchers learned how to copy short stretches of DNA—a process that is referred to as molecular cloning.

Molecular Cloning

Cloning allows for the creation of multiple copies of genes, expression of genes, and study of specific genes. To get the DNA fragment into a bacterial cell in a form that will be copied or expressed, the fragment is first inserted into a plasmid. A plasmid (also called a vector in this context) is a small circular DNA molecule that replicates independently of the chromosomal DNA in bacteria. In cloning, the plasmid molecules can be used to provide a "vehicle" in which to insert a desired DNA fragment. Modified plasmids are usually reintroduced into a bacterial host for replication. As the bacteria divide, they copy their own DNA (including the plasmids). The inserted DNA fragment is copied along with the rest of the bacterial DNA. In a bacterial cell, the fragment of DNA from the human genome (or another organism that is being studied) is referred to as foreign DNA to differentiate it from the DNA of the bacterium (the host DNA).

Plasmids occur naturally in bacterial populations (such as *Escherichia coli*) and have genes that can contribute favorable traits to the organism, such as antibiotic resistance (the ability to be unaffected by antibiotics). Plasmids have been highly engineered as vectors for molecular cloning and for the subsequent large-scale production of important molecules, such as insulin. A valuable characteristic of plasmid vectors is the ease with which a foreign DNA fragment can be introduced. These plasmid vectors contain many short DNA sequences that can be cut with different commonly available restriction enzymes. Restriction enzymes (also called restriction endonucleases) recognize specific DNA sequences and cut them in a predictable manner; they are naturally produced by bacteria as a defense mechanism against foreign DNA. Many restriction enzymes make staggered cuts in the two strands of DNA, such that the cut ends have a 2- to 4-nucleotide single-stranded overhang. The sequence that is recognized by the restriction enzyme is a four- to eight-nucleotide sequence that is a palindrome. Like with a word palindrome, this means the sequence reads the same forward and backward. In most cases, the sequence reads the same forward on one strand and backward on the complementary strand. When a staggered cut is made in a sequence like this, the overhangs are complementary.

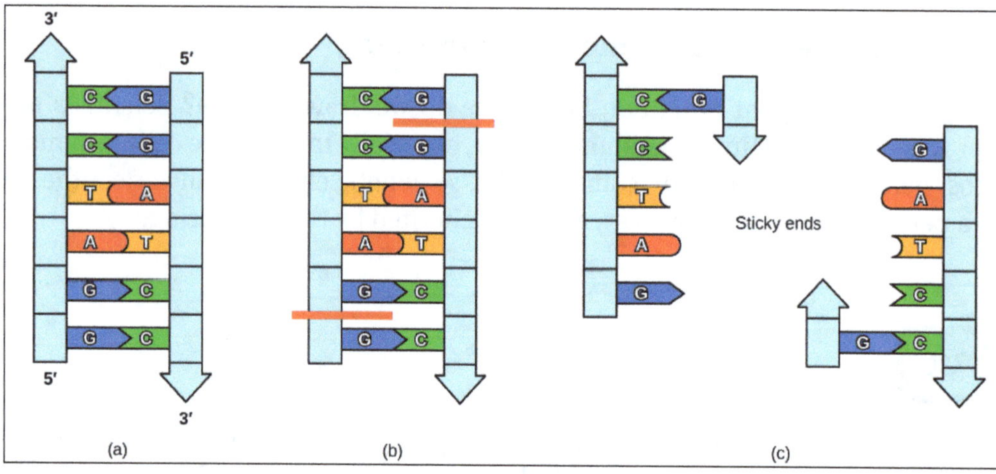

Figure shows (a) six-nucleotide restriction enzyme recognition site, notice that the sequence of six nucleotides reads the same in the 5' to 3' direction on one strand as it does in the 5' to 3' direction on the complementary strand. This is known as a palindrome. (b) The restriction enzyme makes breaks in the DNA strands, and (c) the cut in the DNA results in "sticky ends". Another piece of DNA cut on either end by the same restriction enzyme could attach to these sticky ends and be inserted into the gap made by this cut.

Because these overhangs are capable of coming back together by hydrogen bonding with complementary overhangs on a piece of DNA cut with the same restriction enzyme, these are called "sticky ends." The process of forming hydrogen bonds between complementary sequences on single strands to form double-stranded DNA is called annealing. Addition of an enzyme called DNA ligase, which takes part in DNA replication in cells, permanently joins the DNA fragments when the sticky ends come together. In this way, any DNA fragment can be spliced between the two ends of a plasmid DNA that has been cut with the same restriction enzyme.

Plasmids with foreign DNA inserted into them are called recombinant DNA molecules because they contain new combinations of genetic material. Proteins that are produced from recombinant DNA molecules are called recombinant proteins. Not all recombinant plasmids are capable of expressing genes. Plasmids may also be engineered to express proteins only when stimulated by certain environmental factors, so that scientists can control the expression of the recombinant proteins.

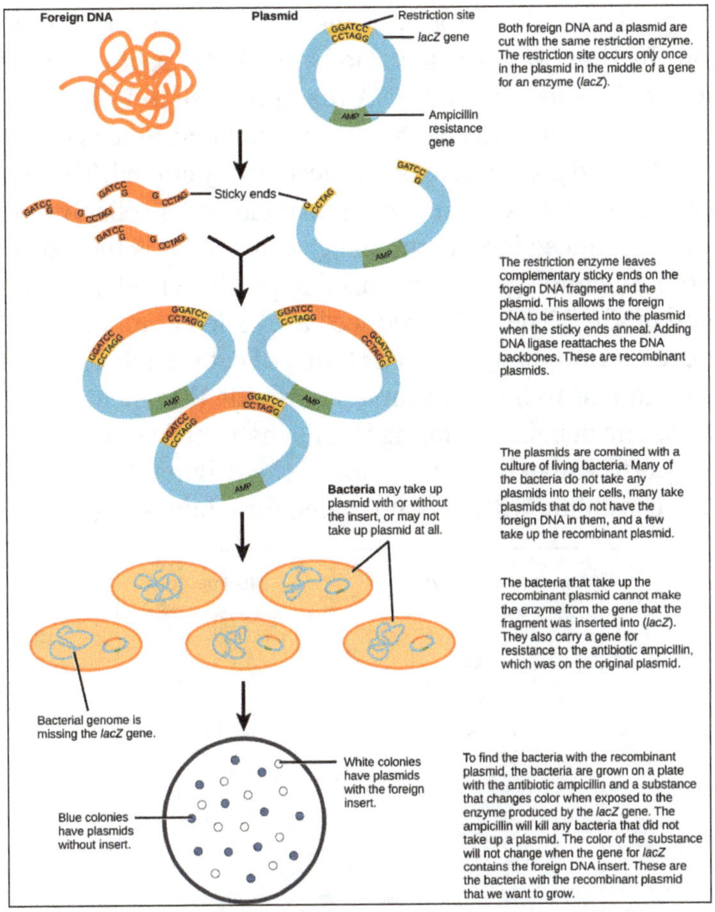

This diagram shows the steps involved in molecular cloning.

Reproductive Cloning

Reproductive cloning is a method used to make a clone or an identical copy of an entire multicellular organism. Most multicellular organisms undergo reproduction by sexual means, which involves the contribution of DNA from two individuals (parents), making it impossible to generate an identical copy or a clone of either parent. Recent advances in biotechnology have made it possible to reproductively clone mammals in the laboratory.

Natural sexual reproduction involves the union, during fertilization, of a sperm and an egg. Each of these gametes is haploid, meaning they contain one set of chromosomes in their nuclei. The resulting cell, or zygote, is then diploid and contains two sets of chromosomes. This cell divides mitotically to produce a multicellular organism. However, the union of just any two cells cannot produce a viable zygote; there are components in the cytoplasm of the egg cell that are essential for the early development of the embryo during its first few cell divisions. Without these provisions, there would be no subsequent development. Therefore, to produce a new individual, both a diploid genetic complement and an egg cytoplasm are required. The approach to producing an artificially cloned individual is to take the egg cell of one individual and to remove the haploid nucleus. Then a diploid nucleus from a body cell of a second individual, the donor, is put into the egg cell. The egg is then stimulated to divide so that development proceeds. This sounds simple, but in fact it takes many attempts before each of the steps is completed successfully.

The first cloned agricultural animal was Dolly, a sheep who was born in 1996. The success rate of reproductive cloning at the time was very low. Dolly lived for six years and died of a lung tumor. There was speculation that because the cell DNA that gave rise to Dolly came from an older individual, the age of the DNA may have affected her life expectancy. Since Dolly, several species of animals (such as horses, bulls, and goats) have been successfully cloned. There have been attempts at producing cloned human embryos as sources of embryonic stem cells. In the procedure, the DNA from an adult human is introduced into a human egg cell, which is then stimulated to divide. The technology is similar to the technology that was used to produce Dolly, but the embryo is never implanted into a surrogate mother. The cells produced are called embryonic stem cells because they have the capacity to develop into many different kinds of cells, such as muscle or nerve cells. The stem cells could be used to research and ultimately provide therapeutic applications, such as replacing damaged tissues. The benefit of cloning in this instance is that the cells used to regenerate new tissues would be a perfect match to the donor of the original DNA. For example, a leukemia patient would not require a sibling with a tissue match for a bone-marrow transplant.

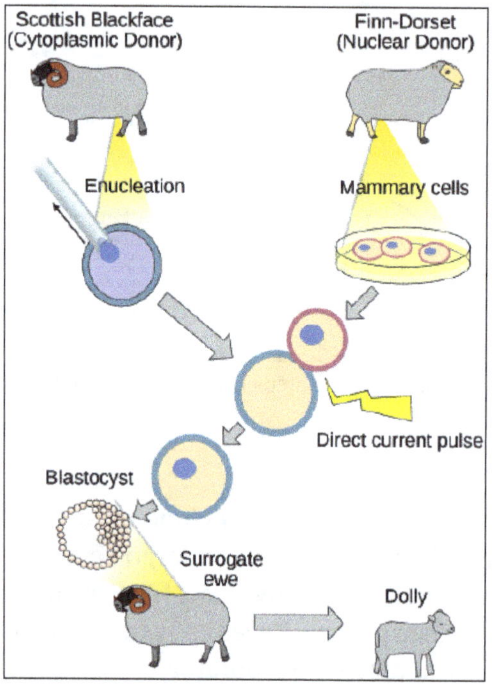

Dolly the sheep was the first agricultural animal to be cloned. To create Dolly, the nucleus was removed from a donor egg cell. The enucleated egg was placed next to the other cell, then they were shocked to fuse. They were shocked again to start division. The cells were allowed to divide for several days until an early embryonic stage was reached, before being implanted in a surrogate mother.

Transgenic Technology

The advent of genetic engineering has enabled us to transform an organism's genetic material in variety of novel ways and thus it has the potential proficiency to address important problems of modern medicine, fulfil the needs of humans. Transgenic technology is the science of intentionally introducing a foreign gene or genetic construct into the genome of a target animal. This in vitro work rapidly progressed and successfully applied into laboratory animal models. This novel technology provides a precise mechanism for the study of specific gene function, facilitating us for a more targeted and proactive approach for the establishment of new animal models for biomedical research, which in turn expanded the applicability of this technology in various segments of health care.

Animal and plant transgenic technology is one of the fastest growing biotechnologies these days. It is used to integrate foreign genes into the animal genome by genetic engineering technology so that foreign genes can be expressed and inherited to the offspring. The efficiency of the transgenesis process and precise control of gene expression are the key limiting factors on preparation of transgenic animals. Numerous variety of transgenic techniques are available, each of which has its own merits and drawbacks and still needs further study to resolve technical and safety issues. With the in-depth research, the transgenic technology will have broad application prospects in the fields of exploration of gene function, genetic improvement of animals, and production of pharmaceutically valuable proteins using animal bioreactor, animal disease models, and organ transplantation and so on. Various definitions for the term transgenic organism exist among the scientific community. A transgenic or genetically modified, organism can be defined as an animal in which there has been a deliberate modification of its genome, the genetic makeup of an organism responsible for inherited characteristics, through a human technological intervention.

Expression Systems used for Transgenic Bioreactor

Genetically modified organisms are considered as versatile bioreactors that are used to produce proteins of several applications. Choosing the right system for foreign gene expression can be particularly important in obtaining biologically active recombinant protein.

1. Bacteria: Bacteria are limited by their poor capacity to synthesize some proteins those which have a complex structure. Some proteins that are abundantly produced in bacteria form aggregates from which makes the isolation process so tedious. Bacteria cannot always fold proteins in an appropriate manner and assemble subunits to form biologically active molecules. Bacteria are limited in their ability to perform the posttranslational protein modifications such as glycosylation, g-carboxylation, phosphorylation and sulphatation, necessary for many targets. Other systems have thus been implemented and some of them proved more or less capable of producing at a low cost, large amount of active proteins with an appropriate biochemical structure.

2. Yeast: Yeast was expected to have some advantages over bacteria. Significantly these cells are eukaryotes have potential ability to synthesize large amount of matured recombinant proteins. Advanced genetic tools have been developed for use in Saccharomyces cerevisiae, Kluyveromyces lactis, Pichia pastoris, Hansenula polymorpha. Interestingly, genetically modified yeast expressing foreign genes coding for enzymes responsible for glycosylation proved to secrete substantial amounts of recombinant proteins having carbohydrates almost similar to those found in human proteins. Nevertheless, yeast cells also lack the ability to produce some of the protein in a biologically active form as compared to their native counterparts. This is mostly due to differences in protein glycosylation. Alteration in glycosylation pattern may affect their in vivo function. Also yeast cells suffers the problem of strain stability, since expression of transgenic proteins and functioning as active biological depends on the genetic background of the host strain.

3. Baculovirus: Baculovirus are the most prominent viruses known to affect the insect population. They are double-stranded, circular, supercoiled DNA molecules in a rod-shaped capsid. More than 500 baculovirus isolates (based on hosts of origin) have been identified, most of which originated in arthropods, particularly insects of the order Lepidoptera. Two of the most common isolates used in foreign gene expression are Autographa californica multiple nuclear polyhedrosis virus (AcMNPV) and Bombyx mori (silkworm) nuclear polyhedrosis virus (BmNPV).The Baculovirus Expression Vector System (BEVS) is one of the most powerful and versatile expression systems available to prepare proteins. The BEVS is a viral derived vector system which has been used to express heterologous genes from many diverse sources in insect cells. The BEVS has several advantages such as expression using this system results in functionally active recombinant protein, expected post-translational modifications, high expression levels, capacity for large inserts, simultaneous expression of multiple genes, ease of purification. Despite these potential advantages, this expression system still suffers several limitations, differences in proteins expression, inefficient secretion from insect cells, improperly folded proteins, proteins that occur as intracellular aggregates, low levels of expression. Many of these troubleshooting limitations have been overcome and succeeded by employing suitable strategies. Another major disadvantage of this method is that expression results in a mixture of recombinant and original parental virus after the initial round of replication. Consequently, the recombinant virus has to be separated from the parental virus by plaque purification, a process that is labour-intensive, technically demanding, and time-consuming.

4. Mammalian Cell Culture: Animal or human cells and particularly CHO cells (Chinese Hamster Ovary) have been used for two decades successfully. The major advantage of these cells is that the proteins secreted in the culture medium are post-translationally modified essentially as their native counterparts. The major drawback of the mammalian cell culture systems is their relatively low production level which consequently leads to limited use of these systems. Although the conditions of animal cell culture were improved, the production of recombinant proteins and their glycosylation may be not stable and reliable.

5. Transgenic Organisms: Transgenic vertebrates and invertebrates have been developed in a variety of species, with both scientific and commercial value including fish, amphibians, birds, mammals, arthropod fruit fly Drosophila melanogaster, nematode Caenorhabditis elegans and plants etc.

Transgenic plants offer interesting opportunity to produce recombinant proteins this technology is relatively easy in plants. A number of enzymes used for research or for diagnosis are currently being produced by transgenic plants at an industrial scale. This system offers several advantages-ease of scaling the process, simplicity in methods to rescue the plant lines allowing a reproducible production of proteins. Transgenic plants also face several serious limitations such as technical difficulty in bringing out accurate post-translational modifications, unwanted and unexpected deleterious immune response in the subjects. In addition, it faces difficulties in purification of protein of interest from leaves in presence of proteases or substances like polyphenols which are not well tolerated by patients. Transgenic proteins obtained from plants may have very little chance to contain pathogens for humans or animals. In addition, using transgenic plants raise ethical problems and one among them is the uncontrolled dissemination of the proteins thus of the antigens when plants are cultured in open fields.

Transgenic animals offer particularly attractive possibilities to prepare recombinant pharmaceutical proteins. The advantages are the low cost production as well as the high quality of the proteins. Compared with other systems, transgenic animals are an attractive platform because they represent powerful tools to address the growing need for therapeutic recombinant proteins. Their ability to produce complex, biologically active proteins in an efficient and economic manner is superior to those of bacteria, yeast, baculo virus based expression vector system, mammalian cells, and transgenic plants. With this system, protein production costs are substantially lower for transgenic animals than for cell culture. Therefore, transgenic animal bioreactors show a financial advantage over other systems. Numerous proteins produced by these systems are now in clinical trials and a biological product, called ATryn (human AT) has been approved for use in human medicine. The generation of transgenic plants, animals in broad aspect including their merits, pitfalls and future perspectives.

Significant Models in Transgenic Technology

1. Mouse: The genetic and physiological similarities between mice and humans have focused considerable attention on rodents as potential models of human health and disease. The advent of genomic manipulation has quickly led to the creation and use of genetically engineered mice as powerful tools for cutting edge studies of human disease research, including the discovery, refinement, and utility of many currently available therapeutic regimes. In vivo studies of gene expression and biochemical pathways often yield very different results relative to ex vivo experimentation using either primary cell cultures or tissue/ organ explants due to the genetic and physiological complexities of a living organism. Thus these models often inadequately represent human subjects. The shortcomings of these model systems have driven the use of the mouse to model human health and disease.

The shared sequence complexity between mice and humans is astounding. Although some physiological and immunological differences exist between mice and humans, genomic sequencing has demonstrated that of ~30,000 genes in both mice and humans only 300 are unique to either species. All mice in a given study are generally are genetically identical, eliminating potential complicating factors such as genetic variation and unique gene polymorphisms. The inter-generational interval of mice is less than three months. Litter sizes are also large enough to permit the rapid expansion of experimental subjects into statistically relevant allies.

2. Swine: The ability to genetically modify pigs has enabled scientists to create pigs that are beneficial to humans in ways that were previously unimaginable. Improvements in the methods to make genetic modifications have opened up the possibilities of introducing transgenes, knock-outs and knock-ins with precision. The benefits to medicine include the production of pharmaceuticals, the provision of organs for xenotransplantation into humans, and the development of models of human diseases. Applications of the modified animals have provided evidence that genetically modified swine have the potential to overcome both physiologic and immunologic barriers that have previously hampered this field.

One of the major advantages of the pig as a potential xenograft donor is the ability of the genome of these animals to be modified by modern techniques of genetic engineering. And also their breeding characteristics, in particular relatively short gestation time and large litter size, combined with the availability of techniques for oocyte manipulation and artificial insemination, have made it possible to carry out both addition of genes (i.e. transgenic animals) and elimination of gene expression (i.e. knockout animals) for selected genes involved in xenograft rejection.

3. Rabbit: Rabbit is the smallest non-rodent laboratory animal model that has been well characterized. It offers certain significant advantages of relatively short gestation time, sexual maturity period, easy of handling during reproduction period that favours the scientists to transform it into transgenic models. Since rabbits are efficient breeders, transferring of exogenous genes in to rabbits and making them to express in its mammary gland is an attractive technique for improving the production of recombinant proteins used in biomedical research. Rabbits are safer than other dairy livestock and poultry because there has been no known serious disease complications that may transmissible through rabbit to humans. In the year 1998, researcher, attempted and succeeded in mammary gland specific expression of Human nerve growth factor beta (hNGF-ß) in transgenic rabbits. A number of recombinant proteins have been produced from rabbit milk, a few of them are a-glucosidase, interleukin-2, tissue plasminogen activator, erythropoietin, lactoferin, chymosin, interferon beta and antithrombin. Research work was also carried out in creating transgenic rabbit models of atherosclerosis, which extends the use of rabbit as transgenic model to a higher level.

4. Caenorhabditis elegans: Following the generation and characterization of chromosomal mutations, the ability to generate transgenic lines in C. elegans revealed the system for the rapid genetic characterization of many diverse biological phenomena. The number of laboratories using the freeliving nematode C. elegans is rapidly growing.The acceptance of this biological model in this field is attributed to its rapid generation time and short life span, easy and inexpensive maintenance, fully sequenced genome, and array of RNAi resources. Additionally, analysis of the C. elegans genome revealed an interesting fact of great similarity between worms and higher vertebrates, which suggests that research in worms could be an important adjunct to studies performed in whole mice. A powerful and important part of worm research is the ability to use transgenic C.elegans to study gene localization and function.

5. Zebra Fish: Fish may present another specific, unique, and unexplored opportunity for use it as transgenic bioreactors for production of important proteins. Zebrafish has been a vital and prominent vertebrate model in a variety of biological disciplines. Extensive information gathered from developmental biology research and genetic research, together with completion of the zebrafish genome project, has placed it in an attractive position for use transgenic model. Currently, GloFish are the only recombinant-DNA animal that has been approved for human use by the U.S. Food and Drug Administration. Many significant researches have been performed worldwide in recent years with Zebra fish transgenic model to benefit the mankind.

6. Avian Species: Avian species have traditionally served as model systems for the study of embryonic development and physiology. Historically they are commercially valuable for manufacturing of influenza vaccine. The chicken embryo has been a significant model of choice for many developmental biologists due to its accessibility and ease of manipulation in its natural environment in the shell. These may be reason for avian species being one of transgenic model in recent years. The most efficient technique involves the use of replication-deficient lentiviral vectors to deliver foreign DNA into the avian germline. The germline can be targeted by delivering DNA to the zygote (single-cell embryo), early embryo, or primordial germ cells (PGCs1). In birds, studies have described the efficient introduction of foreign DNA into the germline by the injection of lentiviral vectors in embryos of freshly laid eggs.

Protein Production Platforms using Transgenic Animals

Recombinant proteins are produced from transgenic animal body fluids. Milk, egg white, blood, urine, insect larvae haemolymph, seminal plasma, and silkworm cocoon derived from transgenic expression models/animals are valuable candidate sources of recombinant protein production at an industrial scale.

In 1985, Hammer and colleagues established the first transgenic livestock animals, including sheep, rabbits, and pigs, in an attempt to develop a way to produce recombinant proteins from these animals. Since then, revolution in the field of producing numerous recombinant proteins from transgenic animals was made as a history. Selection of a suitable method for expressing a recombinant protein is dependent on the characteristics and intended application of the recombinant protein. Presently milk is the most studied system for producing recombinant protein from transgenic organisms. The technology for using the mammary glands of transgenic animals as primary bioreactors has been developed for large transgenic animals, such as cows, goats, sheep, and pigs. In addition, many laboratories and pharmaceutical companies have made efforts to produce a variety of valuable therapeutic proteins using transgenic rabbits. Blood, egg white, seminal plasma, silk gland and urine are other theoretically possible systems, which has long way to go.

1. Mammary gland, the best bioreactor available: The mammary gland has been considered the organ of choice to express valuable recombinant proteins in transgenic volumes. Several scientific reports were documented that foreign transgenic proteins are produced in milk at rates of several grams per litre. Apart from several successes, a certain number failures have occurred for various reasons. Certain drawbacks of this platform are lack of fundamental knowledge in areas such as protein maturation and secretion and adverse effects of bioactive proteins produced in milk on animal's health.

2. Blood: alternative recombinant protein secretion medium: Animal blood, which collects secretions from many tissues, may be used as an alternative source of recombinant proteins. For example, human an AT was obtained at a high level from the serum of transgenic rabbits and human haemoglobin has been produced in a transgenic swine circulatory system.

3. Transgenic eggs: Chickens represent an important alternative with a number of distinct advantages as animal bioreactors. An interesting fact in this system is that a single hen can produce an impressive number of eggs and their eggs are extremely rich sources of protein and unlike milk, egg contents are naturally sterile. Second, transgenic chickens require less time to mature than large livestock such as goats or cows; thus using transgenic chickens would significantly reduce the cost and duration of the development process. Third, infrastructure to generate large numbers of eggs at low cost is already in place and, due to the extensive use of eggs in vaccine production, good manufacturing practices exist for the extraction of biopharmaceuticals from chicken eggs. For these reasons, chicken eggs are an attractive system for the production of therapeutic proteins. However, the egg system has been hindered by certain technical limitations and lack of an efficient transgenesis system in poultry.

4. Urine or Seminal-Fluid-Specific Expression Systems: Urine is an abundant biological fluid already used to prepare proteins such as gonadotropins for pharmaceutical use. It is also possible

for a foreign gene to get expressed in a more appropriate manner in the urothelium. Several works by researchers has indicated that expression of the human growth hormone gene driven by the mouse uroplakin II gene promoter was expressed specifically in the urothelium. Compared with milk, one advantage of using the bladder as a bioreactor is that animals can urinate earlier than they can lactate.

5. Silkworm Cocoon: The silkworm has acquired the ability to synthesize bulk amounts of silk proteins in its silk glands. This natural ability of silkworm was merged with transgenic technology for mass production of useful proteins. Transgenic silkworms have been generated that synthesize pharmaceutically related recombinant proteins in the silk gland and secrete them into the silk cocoon; many recombinant proteins have been produced using this system, particularly over the last decade. The recombinant proteins of transgenic silkworm are suitable for the production of pharmaceutically significant fabrics for medical purposes. The disadvantages are that it is expensive; time consuming to maintain transgenic silkworms and that the use of chemicals to extract the recombinant proteins may affect its original structure that leads to loss of intended function.

Varieties of recombinant proteins have been made through several transgenic animal species. However, current methods of generating transgenic animal are quite unproductive and time consuming, and attempts to improve transgenesis by various methods resulted in limited success.

Procedure to Create Genetically Modified Organisms

Transgenic Animals

Many different genes and their characteristics have now been identified. This opens up new ways of changing genes that may be useful e.g., to cure diseases or introduce genes in to an animal for a variety of reasons. The insertion of a foreign gene (transgene) into an animal is successful only if the gene is inherited by offspring. The success rate for transgenesis is very low and successful transgenic animals need to be cloned or mated. The underlying principle in the production of transgenic animals is the introduction of a foreign gene (transgenes) into an animal. The transgene "must be transmitted through the germ line, so that every cell, including germ cells, of the animal contains the same modified genetic material.

Construction of Transgene

Although DNA structure is essentially the same for all organisms, the details of the genetic code do differ. Genes from one organism cannot just be introduced, unmodified into another and be expected to work correctly. Firstly, the transgene must be made, that contains the desired gene plus some extra DNA that allows the new gene to express appropriately and at the right time in the new animal. The extra DNA segment that should be positioned next to the gene called the promoter sequence that controls the targeted expression is needed to ensure the gene will function correctly and in the correct tissues of the recipient animal. Finally, a poly A or termination sequence must be added to signal that the end of the gene sequence has been reached.

Transgene in the Models

Three basic methods of introducing transgene in to the models are as follows:

1. DNA microinjection: Gene transfer by microinjection is the predominant method used to produce transgenic animals. The mouse was the first animal to undergo successful gene transfer using DNA microinjection. Microinjection technique is carried out by using specialized equipment that include microinjection needles, microscopes of high sophistication made by different companies and micromanipulator systems fully functional either through by air-driven systems or oil-driven hydraulic systems.

In this method, eggs were harvested from super-ovulated animals and fertilised invitro. The eggs must be identified before mitosis occurs. A microtube is used to hold the fertilised egg in place and an extremely fine needle is used to inject a tiny amount of solution containing many copies of the transgene from another member of the same species or from a different species into the pronucleus of a reproductive cell. Then the manipulated cell, which first must be cultured in vitro to develop to a specific embryonic phase, is then transferred to the recipient female. Only a small proportion of the animals born are transgenic and only a proportion of those express the gene well. These successful animals are called founder animals. Since the insertion of DNA results in a random process, transgenic animals are mated to ensure that their offspring acquire the desired transgene. However, the success rate of producing transgenic animals individually by these methods is very low. For example, gene transfer studies revealed that only 0.6% of transgenic pigs were born with a desired gene after 7,000 eggs were injected with a specific transgene.

2. Retrovirus-mediated gene transfer: Retrovirus-mediated expression cloning was developed during mid1990s. The use of retrovirus offer several potential advantages that include technical ease, effectiveness of gene transfer and target cell specificity. This method involves the use of retroviruses as vectors to transfer genetic material into the host cell, the resultant viral DNA after reverse transcription and integration becomes a part of the host cell genome which would be maintained for the life of host cell. This results in a chimera, an organism consisting of tissues or parts of diverse genetic constitution. Chimeras are inbred for as many as 20 generations until homozygous transgenic offspring are born. The method was successfully used in 1974 when a simian virus was inserted into mice embryos, resulting in mice carrying this DNA. Despite of several merits, this gene transfer technology also suffers from certain limitation such as its limited size range (about 10kb) for reverse transcription, which may affect the expression level.

3. Embryonic stem cell-mediated gene transfer: This method is used when it is important to target gene sequences to specific sites in the genome. When cells are in culture it is possible to use appropriate vectors to carry out specific genetic modifications, such as removing or substituting a specific gene. This method involves isolation of totipotent stem cells from embryos. Then the desired gene is inserted into these cells and the embryonic stem cells can be injected in to the blastocyst stage of the embryo, the resulting foetus will be chimera. Unlike the other two methods, which require live transgenic offspring to test for the presence of the desired transgene, this method allows testing for transgenes at the cell stage. Further selective breeding is then required for homozygous expression.

Techniques to Analyse GMOs

The identification and analysis of transgenic animals is first performed on genomic DNA isolated from biopsy tissue of the animal. The analysis establishes the presence of the transgene, the zygosity, whether transgene rearrangement or deletion has occurred and transgene copy number. This can be done by real time quantitative polymerase chain reaction (qPCR) based assays and or southern blotting. The integration of transgene can be visualized microscopically using fluorescence in situ hybridization (FISH). The next stage of analysis involves determining pattern and level of transgene transcription or the absence of expression in animals in which a gene target has been silenced or deleted. Transcription analysis can be performed by Northern blotting, ribonuclease (RNase) protection assays and reverse transcriptase (RT)-PCR. As precise quantitation is possible with Quantitative (RT)-PCR, it is becoming the technique of choice to analyse expression. Finally, characterization includes analysis of the protein product and its level of expression, using any one of the several techniques such as western blotting, ELISA, RIA and immunohistochemistry. The experimental data of these studies then can be correlated with any phenotype that the animals exhibit.

Transgenic Plant

Genetic engineering techniques have enabled us to modify plant genome in variety of novel ways and that have the potential to address important problems of modern agriculture and medicine. Transgenic plant research depends on the availability of procedures for plant transformation. Generally two types of methods exist, the use of Agrobacterium as biological vector for foreign gene transfer and direct gene transfer techniques.

Gene Transfer using Agrobacterium as Biological Vector

The most widely used method for the introduction of new genes in to plants is based on the natural DNA transfer capacity of Agrobacterium tumefaciens. In nature this soil bacterium causes tumour formation, termed as crown gall on a large number of dicotyledons. During infection a part of the Ti- plasmid (T-DNA) of Agrobacterium, is transferred and integrated into the plant genome. This natural capacity made scientist to use this bacterium as a natural biological vector to transfer foreign genes in to plant chromosomes.

Limitations of using Agrobacterium as Biological Vector

Agrobacterium based method is mostly preferable method because of its simplicity and efficiency, but still suffers the limitation in terms of the range of plant species which are approachable to transformation, that is Agrobacterium infects only herbaceous dicotyledonous species most efficiently and it is less effective on monocotyledonous species. However, attempts were made to adapt the relatively low simple and convenient system to transform recalcitrant plant species. Recent work has shown that successful transformation is possible provided specific cell culture procedures, defining inoculation and co-cultivation conditions, some non-host species have now been transformed by this transfer technique, although there are still many plant species for which this method is not usable.

The Agrobacterium transformation method includes two different procedures; the first one is regeneration dependent transformation procedure while the second is not. The procedure has the potential advantage that allows the recovery of uniformly transformed shoots and the selection of such shoots. Still it has several limitations in developing a suitable regeneration method for many plant species. Therefore many efforts have been made to the development of regeneration –independent transformation procedures, such as meristem transformation and in planta transformation techniques.

The shoot apex has been used in meristem transformation as an attractive target for transformation since it contains the meristematic cells from which all the aerial parts of the plant are derived. Stable transformation has been achieved through meristem transformation with A.tumefaciens though it is having its own limitations such as it does not always result in germline transformation and transmission of transgenes to the offspring. Oryza sativa is the first cereal species that has been stably transformed via Agrobacterium. In planta methods, involves the infecting the imbibed seeds with Agrobacterium, allowed to grow in to mature plants and finally transformants were identified among the seeds harvested from these plants. Another technique, called floral dip, involves dipping of developing floral tissues into an Agrobacterium suspension are also practiced in recent years.

Direct Gene Transfer

Direct gene transfer transformation methods are species and genotype-independent in terms of DNA delivery, but their efficiency is influenced by the type of target cell, and their utility for the production of transgenic plants in most cases depends on the ease of regeneration from the targeted cells, as most methods operate on cells cultured in vitro. Several methods of direct gene transfer exist such as:

- Particle bombardment.
- Protoplast transformation.
- Tissue electroporation.
- Ultrasound induced transformation.
- Silicon-carbide fibre mediated technique.
- Laser-mediated transformation.
- Microinjection.

Of the above direct gene transfer techniques, particle bombardment and protoplast transformation are most widely used today. Particle bombardment technique uses a specialized particle gun to transfer foreign gene in to target plant and this method satisfies the criteria of technical simplicity and reproducibility. Protoplast transformation includes a variety of modified techniques such as treatment of protoplast with DNA in the presence of polyvalent cations, fusion of protoplasts with liposomes containing foreign DNA. Protoplast transformation can be highly efficient, but demands more complicated cell culture techniques. Other direct gene transfer techniques are also quite simple, but some of them are still suffers certain limitations and they are in the development stage which in future may prove to be more promising technologies in this field.

Transgenic technology has created profound effect upon modern agriculture. Transgenic methods have been employed over the 15 years in a number of important crop plants such as maize, cotton, soybean, paddy and a variety of vegetable crops like tomato, potato, cabbage and lettuce. The major breakthroughs of transgenic plant technology include development of specific characteristic features such as tolerance to insect, disease, herbicide and production of food, enzymes, edible vaccines, antibodies with improved product quality.

Applications of Genetically Modified Organisms

Animal biotechnology has been practiced since the beginning of the domestication of animals. Many of the previously used tools of animal breeding, genetics, and nutrition have played and will continue to play an important role in the selection, propagation, and management of desirable and economically important characteristics in livestock. Yet, improvements are still needed in product composition and production efficiency, especially in growth, disease resistance, and reproduction. Genetically modified (transgenic) livestock will have important roles in producing more and higher quality food derived from livestock. Transgenic animals have potentially extensive application for the improvement of animal production quality, the enhancement of production capacity, and the studies of human disease models and production of biomedical materials. The benefits of transgenic technology to human welfare can be grouped in to the following areas.

Agricultural Applications

Practical applications of transgenics in livestock production include enhanced prolificacy and reproductive performance, increased feed utilization and growth rate, improved carcass composition, improved milk production and/or composition, modification of hair or fiber, and increased disease resistance. Development of transgenic farm animals will allow more flexibility in direct genetic manipulation of livestock. Gene transfer is a relatively rapid way of altering the genome of domestic livestock. The use of these tools will have a great impact toward improving the efficiency of livestock production and animal agriculture in a timely and more cost-effective manner. With ever increasing world population and changing climate conditions, such effective means of increasing food production are needed.

Transgenic technology could provide a means of transferring or increasing nutritionally beneficial traits in animals. For example, enhancing the omega-3 fatty acid in fish consumed by humans may contribute to a decreased occurrence of coronary heart disease. In fact, transgenic pigs that contain elevated levels of omega-3 fatty acids have been produced. Furthermore, transfer of a transgene that elevates the levels of omega-3 fatty acids into pigs may enhance the nutritional quality of pork. Another example is advances in transgenic technology provide the opportunity either to change the composition of milk or to produce entirely novel proteins in milk. The improvement of livestock growth or survivability through the modification of milk composition involves production of transgenic animals that: produce a greater quantity of milk; produce milk of higher nutrient content; or produce milk that contains a beneficial 'nutriceutical' protein. Transgenic technology was efficiently applied to tackle the problem of higher mortality rate of piglets attributed to the insufficient content a-lactalbumin in milk. Transgenic pigs that carries alactalbumin gene of cow produced milk containing an increase concentration of lactose in their milk. Lysozyme, Lysostaphin, Bovine ß and κ - casein, IGF-I, a-lactalbumin, and lactoferin are few examples for mammary expression of transgenic proteins.

Genetic modification of livestock will enhance animal welfare by producing healthier animals. One application of this technology is to treat mastitis, an inflammation of the mammary gland, typically caused by infectious pathogen(s). Mastitis causes decreased milk production. Transgenic dairy cows that secrete lysostaphin into their milk have higher resistance to mastitis due to the protection provided by lysostaphin, which kills the bacteria Staphylococcus aureus, in a dose-dependent manner Pioneering researches have been made to produce genetically modified papaya tree variety that resistance to papaya Ring spot virus (PRSV).

Medicine

1. Xenotransplantation: Swine have been a significant model in biomedical research as they are versatile models for cardiovascular disease, atherosclerosis, cutaneous pharmacology, wound repair, cancer, diabetes, ophthalmological studies and toxicology research, as well as being considered as a potential source of organs for xenotransplantation. Today more than 250,000 people are alive only because of the successful transplantation of an appropriate human organ (allotransplantation). However, progress in organ transplantation technology has led to an acute shortage of appropriate organs, and cadaveric or live organ donation does not meet the demand. To close the growing gap between demand and availability of appropriate human organs, porcine xenografts from domesticated transgenic pigs are considered to be the best alternative. Currently, xenotransplantation is hampered by a pig protein that can cause donor rejection but research is underway to remove the pig protein and replace it with a human protein.

2. Production of Pharmaceuticals: ATryn (recombinant human antithrombin III) is the first transgenic recombinant protein from transgenic animal approved by the United States Food and Drug Administration. Many products such as human alpha antitrypsin, MAbs, cystic fibrosis transmembrane regulator, alpha glucosidase, immunoglobulins, tissue plasminogen activator, interferons, cytokines, insulin, growth hormone and blood anticlotting factors are under developmental phase. In addition, production of antibodies, edible vaccines using transgenic plants have gained remarkable achievement.

3. Transgenic animals in field of toxicology: Transgenic animals are genetically designed animals created by the introduction of DNA coding for specific genes into the germline of an organism and that act as a tool to identify and characterize the expression of targeted genes. In the field of toxicology, transgenic animal models have primarily been used to screen for toxicity and to elucidate mechanisms of toxicity. For example, this technology facilitates the identification of chemically induced mutations and helps in determining the mechanism of such mutations. Transgenic animal models are powerful tools for developing a more detailed understanding on the roles of specific genes in biological pathways and systems. These investigations will permit innovative approaches in understanding the mechanism of gene alteration by the agents at the genome level. Transgenic animal models are well utilized, especially for the screening of mutagenic and carcinogenic potential and for the characterization of mechanisms of action of toxicants.

4. Transgenic animals as disease models: Transgenic animals are extensively used to study in vivo gene function during development, organogenesis and aging in turn it is extensively useful to model human diseases. It also permits the evaluation of therapeutic strategies in models of human disease, as well as the investigation of disease progression in a manner not possible in human subjects.

5. Transgenic mosquitoes in prevention of vector borne disease: Genetically modified mosquitoes have been proposed as an alternative strategy to reduce the heavy burden of malaria. Pioneering research in this field has led the scientist to create transgenic lines of Anopheles stephensi, a natural vector of plasmodium falciparum, which constitutively secrete a catalytic inactive phospholipase A_2 ($mPLA_2$) in to the midgut lumen to interfere with plasmodium ookinete invasion. Thus these transgenic lines may be helpful to impair the development of human malaria parasites. In addition, when fed on malaria-infected blood, transgenic mosquitoes are more fecund than non-transgenic mosquitoes. From the experimental evidences it was suggested that the expression of an anti-plasmodium effector gene gives transgenic mosquitoes a fitness advantage when fed malaria infected blood. These findings have important implications for future applications of transgenic mosquito technology in malaria control.

Industrial Applications

In 2001, scientists at Nexia Biotechnologies (Canada) spliced spider genes into the cells of lactating goats. The goats began to express spider silk along with their milk. By extracting polymer strands from the milk and weaving them into thread, the scientists can create a light, tough, flexible material that could be used in such applications as military uniforms, medical microsutures.

Advances and Recent Developments in Transgenic Technology

Remarkable achievements have been attained during the past decade in the field of transgenic technology especially in the development of gene transfer techniques. Novel techniques includes germline stem cell mediated transfer, gene targeting, RNA interference (RNAi)-mediated gene silencing technology, the induced pluripotent stem cells (iPS) transgenic technology, biolistics method, sperm mediated gene transfer, use of engineered molecular tools called Zinc-Finger Nucleases (ZFNs), and use of non-viral episomal vector for gene transfer. These new transgenic techniques were aimed at increasing accuracy and efficiency of the transgenesis and they can provide a better platform for the study of transgenic animals and promote the development of medical sciences, livestock production, and other fields.

1. RNA interference (RNAi)-mediated gene silencing technology: Gene silencing is a term generally used to describe the "switching off" of a gene by a mechanism other than genetic modification. That is, a gene which would be expressed (turned on) under normal circumstances is switched off by machinery in the cell. Constitutive and conditional knockdown of transgene expression in vivo has also been accomplished through RNA interference (RNAi). RNA interference is the sequence specific gene-silencing process that double-stranded RNA induces degradation of cognate mRNA mediated by 21-23 nucleotide short interfering RNA (siRNA). In invertebrates, long double stranded RNAs are processed into short interfering RNAs (siRNAs) by the Dicer ribonuclease. The (siRNAs) antisense strand serves as a template for the RNA-induced silencing complex (RISC) which recognizes and cleaves the complementary mRNA leading to its rapid degradation and knockdown of protein expression.

2. Biolistics: Transgenic animals also been created through the use of biolistic bombardment. Biolistic bombardment is a simple method to introduce foreign DNA into many organisms, including C. elegans. This technique uses accelerated microprojectiles to transfer DNA or other molecules into

intact tissues and cells. Biolistic bombardment is a less familiar and newer technique that involves generation of transgenic worms by biolistic bombardment with gold particles using the Bio-Rad PDS-1000 system. It relies upon gold particles forming a complex with DNA in the presence of $CaCl_2$. Compared with DNA microinjection into hermaphrodite germline, this protocol has the advantage of not requiring special skills with regards to identifying worm anatomy or performing microinjection, thus biolistic bombardment can be an attractive method for the generation of transgenic animals.

3. Sperm-mediated gene transfer (SMGT): Sperm-mediated gene transfer (SMGT) is also an effective method for producing transgenic animal models. SMGT has been achieved using a variety of methods. For example, in one approach, sperm cell membranes were disrupted by repeated freeze–thaw cycles or by exposure to detergents to facilitate introduction of exogenous DNA. In mice and pigs, to facilitate DNA uptake, the recombinant DNA has also been anchored non- covalently to the sperm head with an antibody that recognizes a sperm surface protein. Others have introduced recombinant DNA into the sperm head by electroporation.

4. Use of lentiviral vectors: While conventional microinjection-based methods for transgenesis have been successful in generating small and large transgenic animals, their relatively low transgenic efficiency has opened the door for alternative approaches. Research in recent year's lentiviral vectors have become a promising new tool for the establishment of transgenic animals and the manipulation of the mammalian genome. Lentiviral vectors have proven highly efficient in generating transgenic rats and mice, farm animals, Japanese quail, marmoset monkeys, and rhesus monkeys. For example, transgenic rat models of Alzheimer's disease have been generated via lentiviral vector infection of rat zygotes. Production of transgenic mice and rats has also been accomplished by retroviral/lentiviral transduction of male germ-line stem cells and pantropic pseudotyped retroviral vectors have been used to prepare transgenic fish and crustaceans.

5. Zinc Finger Nucleases technology: The extensive necessity for technology to manipulate cellular genomes has boosted the development of a new classification of engineered molecular tools called Zinc-Finger Nucleases (ZFNs). This Novel technology has recently emerged that facilitate targeted editing of genomes in diverse systems. Targeted editing includes precise manipulation of gene sequences in their natural chromosomal milieu and also addition of transgenes to specific genomic loci. ZFNs are engineered proteins developed for the purpose of introducing site specific mutations in cellular genomes. This advancement has been facilitated by progresses in engineering targeted nucleases with programmable, site-specific DNA-binding domains, including zinc finger proteins and transcription activator-like effectors (TALEs). These advances are driving new approaches to many forefront areas of biotechnology; one among them is creation of transgenic organisms and studies of genome structure, regulation, and function.

Ethical Issues

Advances in transgenic technology bring new responsibilities for safe use of transgenic organisms for the benefit of humanity and the environment. The crucial objective of risk assessment is to develop safety procedures to assess long term effects on human health and the environment. Genetic engineering techniques are still in the research phase, there are varieties of intended applications for their use. Although it may provide considerable benefits in areas such as biomedical science and food production, the creation and use of genetically engineered animals may raise ethical issues related to animal welfare concerns, invasiveness of procedures, unanticipated welfare concerns and use of this technique for well-being of humans in safe manner. FDA has developed guidelines to assure the safe commercial utilization of recombinant biological products. Safety concerns of bioengineered products are one of the significant issues to be discussed. A vital consideration with animal-derived products is the prevention of transmission of pathogens from animals to humans. This necessitates development of sensitive and reliable screening methods for the various types of target organisms. Insertion of foreign gene in to chromosome of an animal may result in unintended genetic changes to different degrees. In certain instance it may increase the chances of zoonotic disease to occur and may also cause allergic reactions in humans.

The generation of new genetically-engineered animals often involves the sacrifice of some animals and surgical procedures (for example, vasectomy and surgical embryo transfer). However, progress is being made to refine the genetic engineering techniques that are applied to mammals, so that less invasive methods are feasible. For example, use of a technique called non-surgical embryo transfer (NSET) acts in a similar way to artificial insemination.

Large numbers of animals required because many of the embryos that undergo genetic engineering procedures do not survive, and of those that do survive only a small proportion carry the genetic alteration of interest. This means that large numbers of animals are produced to obtain genetically engineered animals that are of scientific value, and this contradicts efforts to minimize animal use. One key refinement and reduction effort is the preservation of genetically engineered animal lines through the freezing of embryos or sperm (cryopreservation). Certain other issue to be addressed is the threat posed by these GMOs in the environment. So progress to be made on this important issue, especially for those genetically engineered species that are intended for life outside the research laboratory, where there may be less careful oversight of animal welfare. Consequently, limits to genetic engineering need to be established using the full breadth of public and expert opinion.

Genetic engineering also brings with it concerns over intellectual property, and patenting of created animals and/or the techniques used to create them. Confidentiality of intellectual property of these techniques may create situations in which there is unnecessary duplication of genetically engineered animal lines, thereby challenging the principle of Reduction. Consequently, even if animal welfare can be satisfactorily secured, key ethical concerns about the genetic engineering of animals may obstruct certain types of genetically engineered animals from reaching their intended commercial application. Given the complexity of views regarding genetic engineering, it is valuable to involve all personnel in discussions about the applications of this technology. Possible risks for human health and the environment, associated with transgenic plants are:

- Development of new allergens from the transgenic proteins expressed, which could trigger allergic reactions in humans.
- Formation of new toxins through unexpected interactions between the transgenic protein product and other endogenous constituents of the organism.
- Indirect promotion of problems with antibiotic resistant pathogens through the dispersion of antibiotic resistance genes used as markers from the genetically modified organism derived food to gut microorganisms.
- Development of new pests due to gene transfer from the transgenic plant to related species.
- The transgenic plant may become a new an invader to natural environment.

- The proteins expressed in transgenic plants may show harmful effects on non-target species.
- Development of resistance may arise from the continuous use of the same agent on the target organism.
- Transgenic plant products may interfere with natural biochemical cycles and affects the natural ecosystem.

International harmonization of regulations and procedures for production, testing and handling transgenic plants will be a major challenge. These challenges can be overcome by obtaining more reliable scientific data for the long-term effects of these organisms to the environment. Certain regulatory agencies works on an accepted ethic of animal use in science, which includes the principles of the Three Rs - Reduction of animal numbers, Refinement of practices and husbandry to minimize pain and distress, and Replacement of animals with non-animal alternatives wherever possible. These issues need to be considered by all stakeholders, including veterinarians and can make a valid contribution to the current debate regarding the creation and use of genetically engineered animals.

Future Perspectives

Use of animals and plants for welfare of humanity has a long history. Many remarkable achievements and sophistication have been on the long way for the improvement of human healthcare sector. One of those is transgenic technology that had a tremendous impact on biomedical research and human welfare. Transgenic technology had revolutionized the production of biopharmaceuticals products with efficiencies far superior than any conventional microbial or cell-culture production systems. In addition they provide valuable experimental models to analyse gene function and regulation as well as facilitating identification of new target genes of therapeutic value. They are also being used to develop and test new therapeutic strategies for human diseases. Therefore, it is reasonable to predict that practical use of transgenic technologies has the potential to yield major discoveries that will improve health and the quality of life in the very near future. The utility of biotechnology is limited only by our knowledge of the genes involved, gene function, and gene product interactions. Transgenic technology continues to hold great promise for the humanity and it will not be realized unless we develop and scrutinize procedures, strategies for risk assessment, for the products produced by these technologies. However, it must be met with sound ethical reasoning by scientist, the biotechnology industry, government and public.

Gene Therapy

Gene therapy, also called gene transfer therapy is the introduction of a normal gene into an individual's genome in order to repair a mutation that causes a genetic disease. When a normal gene is inserted into the nucleus of a mutant cell, the gene most likely will integrate into a chromosomal site different from the defective allele; although that may repair the mutation, a new mutation may result if the normal gene integrates into another functional gene. If the normal gene replaces the mutant allele, there is a chance that the transformed cells will proliferate and produce enough normal gene product for the entire body to be restored to the undiseased phenotype.

Approaches to Gene Therapy

Human gene therapy has been attempted on somatic (body) cells for diseases such as cystic fibrosis, adenosine deaminase deficiency, familial hypercholesterolemia, cancer, and severe combined immunodeficiency (SCID) syndrome. Somatic cells cured by gene therapy may reverse the symptoms of disease in the treated individual, but the modification is not passed on to the next generation. Germline gene therapy aims to place corrected cells inside the germ line (e.g., cells of the ovary or testis). If that is achieved, those cells will undergo meiosis and provide a normal gametic contribution to the next generation. Germline gene therapy has been achieved experimentally in animals but not in humans.

Scientists have also explored the possibility of combining gene therapy with stem cell therapy. In a preliminary test of that approach, scientists collected skin cells from a patient with alpha-1 antitrypsin deficiency (an inherited disorder associated with certain types of lung and liver disease), reprogrammed the cells into stem cells, corrected the causative gene mutation, and then stimulated the cells to mature into liver cells. The reprogrammed, genetically corrected cells functioned normally.

Prerequisites for Gene Therapy

Prerequisites for gene therapy include finding the best delivery system (often a virus, typically referred to as a viral vector) for the gene, demonstrating that the transferred gene can express itself in the host cell, and establishing that the procedure is safe. Few clinical trials of gene therapy in humans have satisfied all those conditions, often because the delivery system fails to reach cells or the genes are not expressed by cells. Improved gene therapy systems are being developed by using nanotechnology. A promising application of that research involves packaging genes into nanoparticles that are targeted to cancer cells, thereby killing cancer cells specifically and leaving healthy cells unharmed.

Ethical and Safety Concerns

Some aspects of gene therapy, including genetic manipulation and selection, research on embryonic tissue, and experimentation on human subjects, have aroused ethical controversy and safety concerns. Some objections to gene therapy are based on the view that humans should not "play God" and interfere in the natural order. On the other hand, others have argued that genetic engineering may be justified where it is consistent with the purposes of God as creator. Some critics are particularly concerned about the safety of germline gene therapy, because any harm caused by such treatment could be passed to successive generations. Benefits, however, would also be passed on indefinitely. There also has been concern that the use of somatic gene therapy may affect germ cells.

Although the successful use of somatic gene therapy has been reported, clinical trials have revealed risks. In 1999 American teenager Jesse Gelsinger died after having taken part in a gene therapy trial. In 2000 researchers in France announced that they had successfully used gene therapy to treat infants who suffered from X-linked SCID (XSCID; an inherited disorder that affects males). The researchers treated 11 patients, two of whom later developed a leukemia-like illness. Those outcomes highlight the difficulties foreseen in the use of viral vectors in somatic gene therapy. Although the viruses that are used as vectors are disabled so that they cannot replicate, patients may suffer an immune response.

Another concern associated with gene therapy is that it represents a form of eugenics, which aims to improve future generations through the selection of desired traits. While some have argued that gene therapy is eugenic, others claim that it is a treatment that can be adopted to avoid disability. To others, such a view of gene therapy legitimates the so-called medical model of disability (in which disability is seen as an individual problem to be fixed with medicine) and raises peoples' hopes for new treatments that may never materialize.

Gene Targeting

Gene targeting is a method for modifying the structure of a specific gene without removing it from its natural environment in the chromosome in a living cell. This process involves the construction of a piece of DNA, known as a gene targeting vector , which is then introduced into the cell where it replaces or modifies the normal chromosomal gene through the process of homologous recombination.

The Homologous Recombination Process

Homologous recombination is a process that occurs within the chromosome and which allows one piece of DNA to be exchanged for another piece. It is a cellular mechanism that is probably part of the normal process cells use to repair breaks in their chromosomes. Homologous recombination requires that the pieces of DNA undergoing recombination be almost identical (homologous) in sequence. In addition, sequences on either side of the target should be identical, to promote more efficient targeting and recombination. By constructing a sequence that is homologous to a target sequence (such as a gene), laboratory researchers can replace one of the cell's own copies of a particular gene with a copy that has been altered in some way. It is also possible to replace only a part of a gene, such as one portion of its protein coding region. This permits the introduction of a mutation into specific cellular genes, which can either stop the gene functioning altogether (called a "knock out ") or can mimic changes to genes that have been implicated in human diseases. The ability to target DNA constructs to particular locations in chromosomes is a very powerful tool because it allows the modification of more or less any gene of interest, in more or less any way desired.

Homologous recombination of a DNA vector into a gene of interest can be done in almost any cell type but occurs at a very low frequency, and it is therefore important to detect the few cells that have taken up the gene. Gene targeting vectors are designed with this in mind. The simplest strategy is to include an antibiotic resistance gene on the vector, which interrupts the sequence homologous to the gene of interest and thus makes the inserted gene nonfunctional. This "selectable marker" gene makes the cells that possess it resistant to antibiotics, and can then be used to eliminate cells that are not genetically modified. An example of a selectable marker that is commonly used for this purpose is the puromycin-N-acetyl-transferase (pac) gene, which confers resistance to the antibiotic puromycin, a drug that inhibits the function of ribosomes. After the introduction of the DNA construct, the cells are cultured with puromycin in the medium. This allows the selection of single cells that have incorporated the DNA construct into their own chromosomes. Cells lacking the pac gene will die in a culture medium containing puromycin. Once the puromycin resistant cells have been expanded into cell lines, the DNA of these cells can then be analyzed to select out a subset of the cells in which the introduced construct has integrated into the correct (target) gene.

For reasons that are not yet fully understood, the rate of homologous recombination in mouse embryonic stem (ES) cells is substantially higher than that of most other cells. Once a clone of ES cells with the correct targeting event has been identified, these cells can be used to introduced into the mouse via the process of blastocyst injection, which allows the study of gene function in the bodies of living, intact animals. Until very recently mice were the only organisms in which it has been possible to introduce targeted mutations into the germ line. The development of nuclear transfer (moving the nucleus from one cell to another), however, has allowed gene targeting to be done in other mammalian species, such as sheep and pigs.

Adding or Deleting Genetic Material

As well as mutating or knocking out specific genes, gene targeting allows the introduction of novel pieces of DNA into a specific chromosomal location (this is often termed a "knock-in"). This allows researchers to examine the function of a gene in a variety of ways. For example, it is possible to examine where in the animal the gene is normally expressed by insertion (knock-in) of a fluorescent protein (such as green fluorescent protein, GFP) into the gene so that the cells expressing the gene begin to glow. In addition to changing single genes it is also possible to remove or alter large pieces of chromosomes.

Technologies also now exist that allow genes to be removed not just in a whole animal, but in a subset of cells or in a particular tissue. This can be achieved by modifying the vector to include target sites (termed loxP sites) for an enzyme called Cre recombinase. When the Cre enzyme is present in a mouse cell in which the target gene is surrounded by loxP sites, it will cut this gene out of the chromosome. This allows the function of this gene, which may be required for the mouse to normally develop, to be analyzed in a particular cell type or tissue where only the Cre recombinase is expressed.

Therapeutic Potential of Gene Targeting

It is hoped that gene targeting may eventually become useful in treating some human genetic disorders such as hemophilia and Duchenne muscular dystrophy. Treating human disease by the types of genetic approaches mentioned above is termed "gene therapy." This could, in principle, be achieved by replacing the defective gene with a normal copy of the gene in the affected cells of an individual undergoing treatment. In order to make this potential treatment effective it will be necessary to develop technologies that increase the frequency with which targeting occurs. This is currently the subject of much research.

The development of nuclear transfer technology also has opened up the possible alternative method of using homologous recombination for gene therapy based on cell transfer. Gene targeting would be used to replace the defective genes in selected somatic cells in culture, and their nuclei could then be transferred into stem cells. The stem cells can then be differentiated into the affected cell type (for example, into bone marrow cells for hemophilia) and these cells could then be transplanted to patients.

Making the Targeting Construct

The design and construction of the DNA to be introduced into cells(targeting construct) are critical and depend on the type of mutation required. Gene targeting relies on homologous recombination between the target gene and homologous DNA in the targeting construct (the exact mechanism of homologous recombination during gene targeting is unclear). Targeting constructs therefore carry a stretch of DNA homologous to a region in the target locus, and these can undergo homologous recombination with the target locus in two ways. Replacement events involve two crossovers and result in a stretch of the homology in the target locus being replaced by that from the targeting construct. Insertion events involve a single crossover and result in insertion of the entire construct into the target locus and duplication of the region of homology. To allow selection for these events, targeting constructs usually include a positively selectable marker gene (+SMG, e.g., the neomycin phosphotransferase gene, *neo*) conferring resistance to a drug (e.g., G418). For replacement events, the +SMG must be positioned within the region of homology, but this is not true for insertion events.

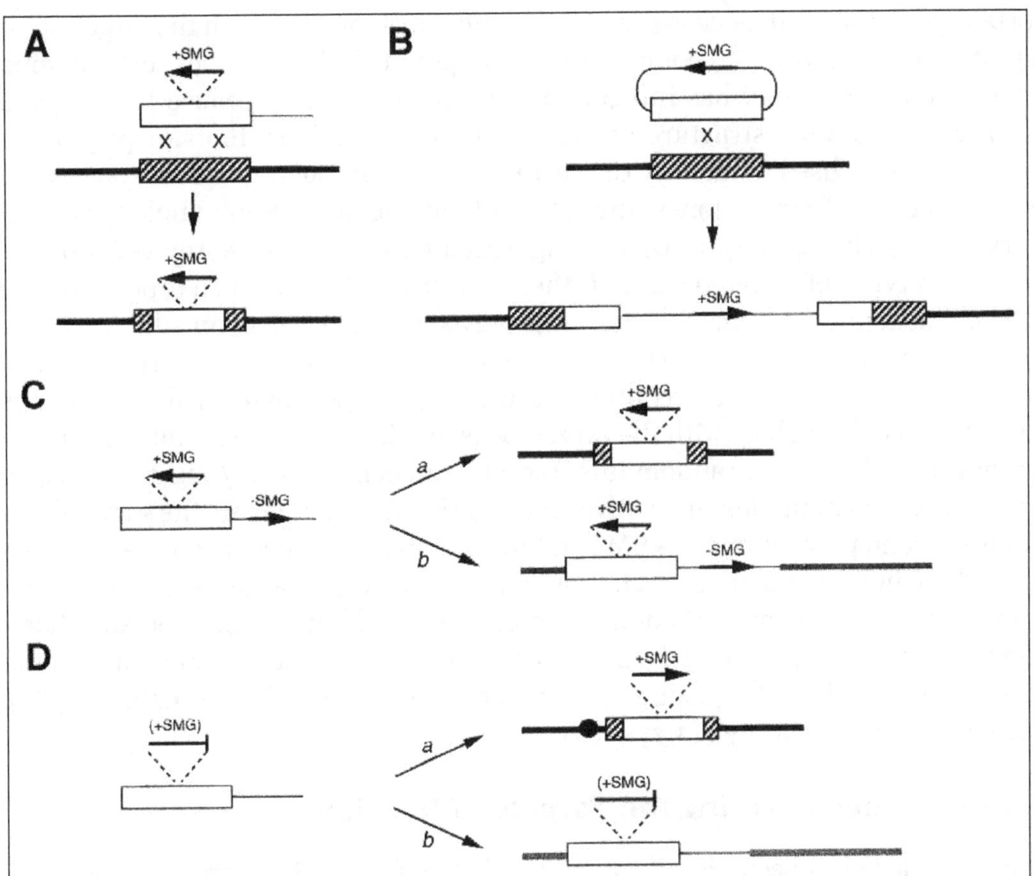

Figure shows basic targeting events. Regions of homology between targeting construct and target locus are shown as white and hatched boxes, respectively. Nonhomologous regions in the targeting construct and target locus are shown as thick and thin black lines, respectively. DNA at a random locus is represented by a thick grey line. Expressed positively or negatively SMG are shown by arrows. (A) A replacement event. (B) An insertion event. (C) A positive/negative targeting construct (left) will survive positive negative selection after homologous integration (a) but not after random integration (b). (D) The inactive SMG in a promoterless targeting construct (left) becomes active after homologous integration (a) but not after random integration (b). The black circle represents the promoter in the target locus.

The efficiency of homologous recombination increases exponentially with the length of homology, although it seems to plateau at about 15 kb, and is impaired by mismatches within the homology. It is therefore best to use homologous DNA that is isogenic with the target gene and to use as much as conveniently possible (usually between 5 and 10 kb).

Targeting the Gene

Targeting constructs are often introduced into the target cells by electroporation, although other transfection methods can be used. Stably transfected clones are usually selected on the basis of a +SMG in the targeting construct and appear as colonies after several days of growth in selective medium.

An important potential bottleneck is the identification of clones in which the targeting construct has integrated by homologous recombination as opposed to those clones, usually much more frequent, where the construct has integrated at a random locus. Because the targeted clones contain a predictable novel structure at the target locus, Southern blots or polymerase chain reaction assays can be used to identify the correct clones. Although these assays can be streamlined for the analysis of many clones, the ratio of homologous to nonhomologous integrations can be very low (e.g., less than 1/1000), making screening very arduous. Any selection procedure that favors the survival of targeted clones is therefore desirable. Two such procedures have been devised. The first applies to replacement events only and makes use of an additional, negatively selectable marker gene (-SMG; e.g., HSVtk, the thymidine kinase gene of herpes simplex virus), which kills cells grown in medium containing drug (e.g., ganciclovir). This gene is positioned outside the region of homology with the target locus and is therefore lost during a replacement event but not (usually) during random integration. For such "positive/negative" targeting constructs, therefore, a combination of positive and negative selection favors the survival of targeted clones. In the second procedure, the +SMG in the targeting construct is deprived of its promoter and is embedded in a promoterless region of homology with the target gene. The promoterless marker remains silent after most random integration events, but is expressed after homologous integration as part of a transcript driven by the target gene promoter. Depending on the nature of the marker gene and how it is positioned within the homology, this "promoter-trap" approach can be implemented in a variety of ways.

Generation of Mice Carrying the Targeted Mutation

Once a targeted clone has been identified and verified by further Southern analysis, it is injected into a host blastocyst derived from a strain of mouse distinguishable (on the basis of coat color) from the strain from which the ES cells were derived. The chimeric blastocyst is implanted in a foster mother and brought to term. The extent to which the targeted ES cells have contributed to the tissues of the offspring can be estimated from the degree of coat color chimerism. Chimeric animals are then mated with normal mice and, if the targeted ES cells have contributed to the germ line, generate some offspring in which all tissues are heterozygous for the targeted mutation. To aid identification of the desired offspring, "normal" mice in these crosses are usually a strain with a different coat color from the strain from which the ES cells were derived. The mice from these matings and their offspring are therefore of mixed genetic background, and this can complicate certain studies, the analysis of graft rejection in particular. To avoid a time-consuming program of inbreeding, it may be worth mating the chimeras with animals isogenic to the ES cells, using analyses of tail biopsy DNA to identify offspring carrying the targeted allele. Once identified, heterozygous animals can be sib-mated to generate embryos homozygous for the targeted mutation and, depending on the effects of the mutation; these will develop into homozygous mice.

Gene Targeting in Somatic Cells

Although initial gene targeting experiments were carried out in cell lines derived from somatic tissues efforts were quickly diverted to gene targeting in mouse ES cells. To date the great majority of gene targeting experiments have been carried out in ES cells. There are two main reasons for this. First, for many target genes, for instance those involved in complex, multicellular processes such as development and immunity, an informative phenotype is not expected to be observed in cell culture. Second, even for (autosomal) genes where targeting might produce a phenotype observable in cell culture, the generation of cells homozygous for a targeted mutation requires two successive gene targeting experiments. Despite these considerations, several somatic cell lines homozygous for targeted mutations have been generated, and experiments of this kind promise to be useful for the genetic analysis of many problems of cell biology.

It is the potential of gene targeting in somatic stem cells, however, that is of particular relevance to transplantation. If somatic stem cells could be grown and genetically manipulated without losing their multipotency and self-renewing properties, as is currently possible with ES cells, the scope for autologous transplantation therapies involving gene targeting would be impressive indeed. At present, however, our knowledge of true somatic stem cells, and how to grow and manipulate them in culture, is severely limiting. Major improvement in this and/or the efficiency of gene targeting itself will be required before the use of gene-targeted cells in autologous transplantation can be seriously contemplated.

Types of Targeted Modification

Gene Disruption

Most gene silencing (knockout) experiments make use of targeting constructs designed for a replacement event. Contiguous arms of homology may be chosen such that the +SMG disrupts a key region of the gene. Alternatively, non-contiguous arms of homology can be used so that the intervening region of the target gene, usually chosen to include a key exon, is deleted by homologous recombination. In either case, the possibility that a viable transcript can still be produced from the targeted gene must be considered. For instance, a viable transcript may be produced via exon-skipping, reflecting either a natural alternative splicing mechanism or a novel mechanism resulting from the targeting event. In the former case, targeting can be used to distinguish the functions of alternately spliced transcripts. For example, we have recently developed mice in which the short or IgV form of CD80 is still produced after the splicing out of exon 3 in which the selectable gene has been inserted, but which do not express full length CD80 (Bugeon, Hargreaves, Rahemtulla, Porter, and Dallman, unpublished data).

Insertion constructs can also be used to disrupt a target gene. The duplication resulting from such insertion events can sometimes give rise, via exon skipping or activation of cryptic promoters, to viable transcripts. Such duplications are also inherently unstable, losing the targeting construct by excision, i.e., by an intrachromosomal, homologous recombination event. For these reasons, and also because they are not compatible with the positive/negative selection procedure, insertion constructs have not been as popular as replacement constructs in gene disruption experiments.

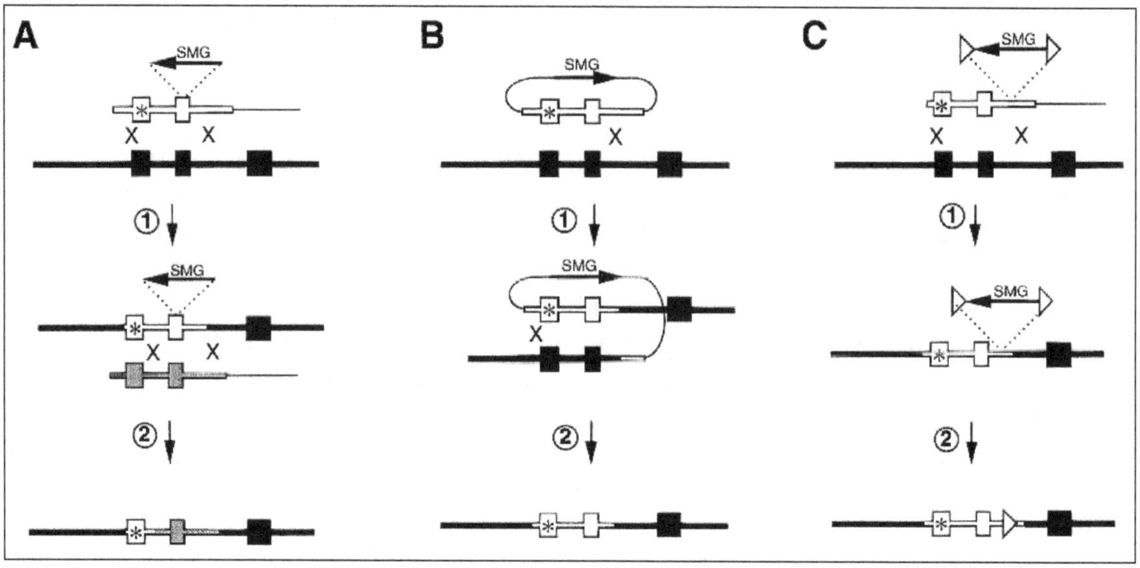

Targeting strategies for the disruption and/or introduction of subtle mutations (asterisks) into a target gene consisting of three exons (black boxes). The target gene DNA is shown in black and homologous DNA in the targeting construct is shown in white (or stippled). Arrows and white arrowheads show positions of SMG and *loxP* sites, respectively. Thin black lines represent non-homologous regions of DNA in the targeting construct.(A) Disruption of target locus by a replacement targeting construct (step 1) followed by removal of the SMG by a second replacement targeting construct(step 2). (B) Disruption of target locus by an insertion targeting construct (step 1) followed by removal of the SMG by excision of the targeting construct (step 2). (C) Disruption of target locus by a replacement construct whose SMG is flanked by *loxP* sequences (step 1) followed by removal of the SMG by Cre-mediated site-specific recombination.

Even if gene disruption is successful in completely abolishing expression of the target gene, an informative and/or reliable phenotype cannot be guaranteed. There are now several examples, in which different phenotypes were obtained after inactivation of the same target gene with different constructs. Discrepancies of this kind probably reflect differences between targeting constructs in their affects on the expression of other genes in the vicinity of the target gene. Such neighborhood effects may reflect the introduction of new*cis*- acting DNA elements (e.g., the promoter/enhancer driving the SMG) or the deletion/disruption of uncharacterized long-range*cis*- acting elements in the target gene. Phenotypes resulting from targeted disruptions of the kind discussed so far must therefore be interpreted with caution and, where possible, in the light of knowledge of the surrounding genes and their control elements.

Subtle Mutations

There are various reasons for wanting to introduce small, even single base pair, mutations into a target gene, without any other accompanying mutations. First, this would allow the role of specific residues within the target gene product to be assessed. Second, rather than being used to introduce mutations, this approach could be used in somatic stem cells to repair the mutations that underlie genetic disorders. Third, subtle (e.g., nonsense) mutations can be used to silence or truncate target genes. Gene silencing by subtle mutation avoids the complications of neighborhood effects associated with gene disruption, allowing phenotypes to be interpreted more reliably.

Small mutations can be included in targeting constructs and will be incorporated into the target gene along with the SMG in a proportion of targeted clones. There are three ways to remove the SMG, all of which rely on a second recombination event that follows homologous integration of the targeting construct. If the first targeting event is with a replacement construct, a second replacement construct can be designed to remove the selectable marker, leaving the subtle mutation in the target locus. In a second approach, sometimes referred to as "in-out" or "hit and run, the subtle mutation is first introduced into the target locus on an insertion construct. After insertion, insertion constructs can excise by intrachromosomal homologous recombination. Because the second crossover can occur anywhere in the region of homology, it is possible for excision to leave the subtle mutation in the target locus. In a third approach, the selectable marker in a replacement construct is first flanked by sites (*loxP*) for the bacteriophage site-specific recombinase Cre. Cells that have been targeted with this construct are then electroporated in the presence of Cre recombinase (or a plasmid that expresses the Cre gene). Transient Cre recombinase activity in the nucleus then catalyses sitespecific recombination between the *loxP* sites, resulting in excision of the intervening DNA. This leaves behind the mutation of interest and a single copy of the 34bp *loxP* sequence. Clearly this strategy requires the *loxP*-flanked SMG to be positioned carefully, so that the single *loxP* site remaining after step 2 does not affect the function of the target gene. An alternative to the Cre/*loxP* system uses the yeast site-specific recombinase FLP and its target sequence FRT.

All three of the approaches illustrated in figure involve loss of the unwanted DNA by a second recombination step (step 2). To select for step 2, a -SMG must be included in the DNA to be lost, and yet a positively selectable marker is also required for step 1. These requirements can be met by using a tandem pair of SMG (e.g., neo and HSVtk), one of which is positively selectable, the other negatively selectable. Alternatively, a single marker gene (e.g., the guanine phosphoribosyl transferase gene) can be used if it is both positively selectable (e.g., in mycophenolic acid) and negatively selectable (in 6-thioguanine).

Conditional Mutations

Many mice homozygous for a targeted mutation are not viable and die in utero or as young animals. Although this may reveal a role for the target gene in development, it is possible that the same gene has another role, perhaps in a different tissue or at a later stage in development. In such situations it would be valuable either to postpone mutation of the target gene until a particular developmental stage or to restrict it to a certain tissue type.

On the other hand, there are many knockout mice that have no obvious phenotype or for which the phenotype is surprisingly mild given the apparent importance of the target gene. A null or surprisingly mild phenotype could of course be explained if the target gene has not been fully inactivated or if the mouse has not been exposed to an appropriate environmental challenge. However, there is also the possibility of genetic redundancy. It is important to distinguish two types of redundancy that we will call simple and complex. In simple redundancy, two or more genes perform a similar role in normal animals, so that after inactivation of one gene, function is maintained by the other-the second gene "covers" for the deficiency in the target gene. Complex redundancy implies a degree of plasticity in the developmental program such that, if the normal program is blocked by a targeted mutation, an alternative pathway is taken that does not require the target gene. An example of such complex redundancy is observed in CD4$^{-/-}$, CD8$^{-/-}$ mice in which a population of CD4^{-} 8^{-} T cells emerges that can function at least partly like their CD4^{+} or CD8^{+} counterparts in normal mice. In simple redundancy, it makes no difference when the target gene is inactivated: there will be no phenotype. This is not the case for complex redundancy, in which inactivation of the target gene late in development should not allow the developmental program time to compensate, and a more pronounced phenotype will emerge. Here again is a situation in which it would be useful to postpone inactivation until a particular stage in development.

Temporal or tissue-specific control of gene inactivation can be achieved by use of the Cre/*loxP* system. Gene targeting is used to introduce two *loxP* sites into the target locus. The positions of the *loxP* sites must be chosen carefully so that they do not affect expression of the target gene until Cre recombinase is expressed and results in deletion of DNA between the *loxP* sites. Spatially or temporally controlled gene disruption can thus be achieved if the gene for the Cre recombinase is driven by a tissue-specific or inducible promoter. It is not necessary to perform all these genetic manipulations in a single cell-line: targeted mice carrying the target gene with two *loxP* sites can be crossed with mice transgenic for a tissue-specific or regulatable Cre gene.

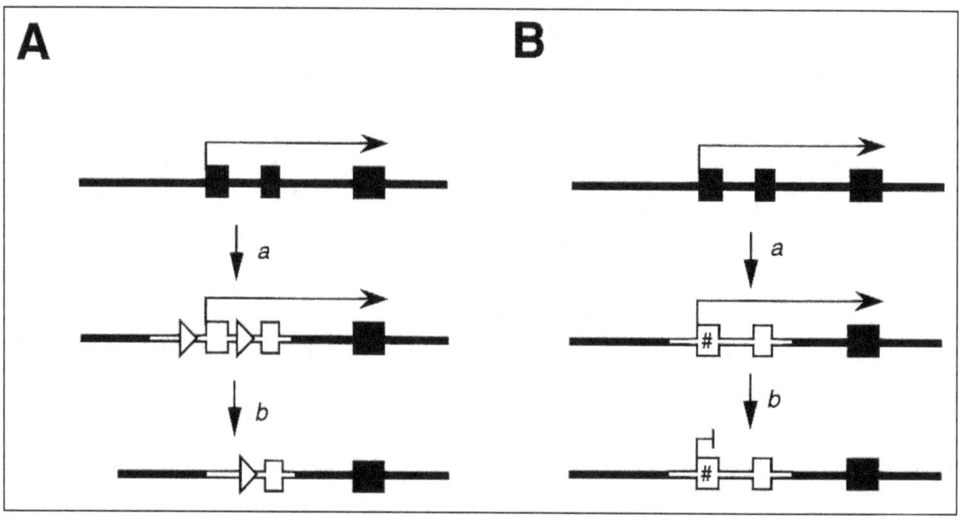

Figure shows targeting strategies for conditional gene silencing. DNA is represented as in figure. Arrows represent active transcription of the target gene. (A) Gene targeting is used to introduce two loxP sites into the target locus (a). This can be achieved using the strategy shown in figure if the subtle mutation is another loxP site. Mice targeted as in (a) are then crossed with mice transgenic for a Cre gene that is expressed from a tissue-specific or inducible promoter. Thus Cre-mediated silencing of the target gene (b) occurs only in a specific tissue type or after Cre expression has been experimentally induced. (B) A cis-acting transcriptional regulatory sequence (e.g., lacO, tetO) is introduced into the target gene (a). This can be achieved by one of the three strategies shown in figure if the subtle mutation is thecis- acting sequence. Expression of the gene for an inducer-responsive transcriptional repressor (e.g., lac repressor) or activator (e.g., tetracycline transactivator), whose target is thecis- acting sequence, places the target gene under the control of the inducer (e.g., isopropyl-β-D-thiogalactoside or tetracycline). Removal(or addition) of inducer results in silencing of the target gene (b).

In another approach to conditional gene inactivation, expression of the target gene itself is made inducible. Thus, gene targeting is used to place a cis-acting element within or near the target gene. Expression of the gene for a regulatable transcriptional activator or repressor that interacts with the *cis*-acting sequence then places the target gene under the control of the inducer. Initially developed in human cells using the isopropyl β-D-thiogalactoside-regulatable *lac* repressor and its recognition sequence *lacO,* this approach is adaptable for use in ES cells using, if necessary, another regulatory system such as that based on the tetracycline transactivator.

Chromosome Engineering

Gene targeting can be used to make gross but defined chromosomal modifications. Chromosomal truncation can be achieved by targeting a stretch of repetitive telomeric DNA to the locus of interest where it can form a new telomere. Large chromosomal deletions or translocations can be achieved by Cre-mediated recombination after targeting of *loxP* sequences to a pair of appropriate loci in the genome, on the same chromosome for a deletion or on different chromosomes for a translocation. Chromosome engineering of this kind may be useful to study the effects of inactivating whole families of genes located at a single locus, e.g., major histocompatibility complex (MHC). These methods may also be used to model or even to repair, natural chromosomal rear-rangements, to construct mammalian artificial chromosomes, and to generate defined genomic regions that have lost heterozygosity with a view to identifying novel recessive mutations.

Gene Targeting in the Analysis of Graft Rejection

Mice with targeted mutations of several genes involved in immunity have now been used as either donor or recipient in transplantation experiments. Examples of such studies are detailed below where they can usefully illustrate some of the points made in the preceding section. The great majority of the experiments have been performed using mice with no expression of the target gene and as yet there has been little work done on mice with subtle or conditional mutations. As a result of this, the interpretation of many of these experiments has been difficult and confounded by the problems of genetic redundancy to which we have already alluded. The use of mice with spatially or temporally controlled mutations will undoubtedly make a further major contribution to knowledge in this area.

Initial hopes that experiments using knockout mice would not only provide definitive answers to several unresolved issues in transplantation but also indicate preferred targets for immune intervention have often been frustrated by the tremendous elasticity of the immune system. Perhaps the most important conclusion that can be drawn from the experiments described below is that transplants may be rejected by many different routes-the removal or negation of one effector mechanism may simply allow another to evolve and cause graft rejection.

Class I, Class II, and CD4 or CD8 Knockout Mice

Mice with disrupted expression of beta 2 microglobulin (class $I^{-/-}$ mice) or MHC class II genes (class $II^{-/-}$ mice) have been generated and used as either donor or recipient of transplants. In many of these experiments, a lack of class I or class II antigens alone on donor tissue has little effect on graft surviva. However, in other experiments, graft survival may be prolonged or permanent when the donor tissue lacks class I only, class II only, or both class I and II antigens. It is clear from these experiments that this variability in results is due both to the type of graft transplanted (i.e., skin versus vascularized) and the strain combinations used. The rejection of class $II^{-/-}$ grafts by normal recipients in a $CD4^{+}$ cell-dependent fashion has, however, clearly implicated indirect antigen presentation in the initiation of graft rejection.

Recipients lacking class I or II antigens are severely depleted of $CD8^{+}$ or $CD4^{+}$ cells, respectively, as are those mice in which either cd8 or cd4 has been disrupted. Importantly, however, despite an apparent depletion of $CD8^{+}$ cells in class $I^{-/-}$ mice, $CD8^{+}$ cytotoxic T lymphocytes (CTLs) can be generated in large numbers after transplantation, and may indeed be involved in the rejection process. Varying results are obtained in experiments which use MHC, CD4, or CD8 knockout recipient mice to assess the activity of CD4 and CD8 cells in graft rejection, an outcome that is perhaps not surprising given the results of other work in which the role of these two T cell subsets has been investigated. Although the central role for $CD4^{+}$ cells in the initiation of graft rejection has been confirmed, depending on the strain combination used and the type of graft, $CD8^{+}$ cells assume a greater or lesser importance.

The interpretation of these apparently straightforward experiments is confounded not only by the ususual features of the immune systems of mice carrying disrupted forms of these genes (for instance the presence of functional $CD4^{-}CD8^{-}$ T cells in mice disrupted at the *cd4* and *cd8* loci and the emergence of $CD8^{+}$ cells in class $I^{-/-}$ mice) but also by the suggestion that grafts from class $I^{-/-}$ mice may be reconstituted in their expression of class I by serum beta 2 microglobulin in the recipient. Further, the potential for activation of natural killer cell killing by class $I^{-/-}$ targets may introduce another variable into these experiments.

Knockout of Cytotoxic Cell Mediators

CTLs are generated in response to a transplant and can be involved in tissue damage. These cells are able to kill their targets through the elaboration of perforins and granzymes, activation of the Fas death pathway, or secretion of the cytokine tumor necrosis factor-α. The involvement of CTLs in graft rejection has been called into question after the finding that mice deficient in perforin are able to reject both tumo, skin, and organ grafts, even when the grafts are resistant to Fas and tumor necrosis factor-α-mediated killing, however, grafts mismatched only at the class I locus are rejected more slowly in perforin knockout mice, indicating that, for class I incompatible grafts at least, cytotoxic cells may play an important role in rejection.

CD18 Knockout Mice

Mice reported by researcher are examples of animals in which disruption of the target locus, in this case cd18, has been performed by insertion, and these mice illustrate a potential pitfall with the use of such targeting constructs. The insertion event resulted in a weakly expressed rather than null allele due to the use of a cryptic promoter in the targeting construct. Nevertheless, such mice have impaired integrin-dependent adhesion and show poor inflammatory responses. Cardiac graft rejection in CD18 knockout mice is delayed, but only slightly so even in the rather weak transplant model used.

Cytokine Knockout Mice

Cytokine knockout mice often exhibit a relatively mild phenotype with little apparent disruption of the immune system. This mild phenotype may result from genetic redundancy, a feature that seems to be particularly prominent in the cytokine network. The redundancy results, at least in part, from the common usage of receptor chains by more than one cytokine. A good example of this is the interleukin (IL)-2 receptor gamma chain that is used not only by IL-2, but also by IL-4, IL-7, IL-9, and IL-15. Mice lacking expression of the γ chain show dramatic defects within the immune system and symptoms similar to those of patients with X-linked immunodeficiency, a condition resulting from mutations in the γ chain.

Although the expression of T helper (Th)1 type cytokines is often associated with acute graft rejection, mice disrupted in expression of either IL-2 or interferon (IFN)-γ reject their grafts with fairly normal kinetics. In IL-2 knockout mice, a strong IFN-γ response is still observed as is the presence of IL-15 transcripts. IL-2 is not absolutely required for graft rejection in such mice, but suggest that this may be due to genetic redundancy in which another cytokine, for instance IL-15, can fulfil the role of IL-2.

It has been suggested by several groups that a Th2-dominated response associates with, and indeed may be responsible for, tolerance induction. Because the presence of IL-4 is an important factor in driving a Th2 response, two groups have used IL-4 knockout mice to investigate this suggestion further. These groups found that induction of tolerance was not impaired in IL-4 knockout mice and, perhaps surprisingly, that it may be easier to induce tolerance in IL-4$^{-/-}$ mice than in heterozygous IL-4$^{+/-}$ mice. This has been interpreted to suggest that in fact IL-4 and/or a Th2 response do not positively contribute to tolerance induction, but rather that they provide a weaker barrier to successful transplantation than do IL-2, IFN-γ, or a Th1 response. This is certainly consistent with the data and views of other workers.

Potential Applications for Gene Targeting in Clinical Transplantation

Xenotransplantation

There has been tremendous interest in the development of xenografts for use in clinical organ transplantation, and indeed the development of pigs carrying human complement regulatory transgenes has been widely publicized. The development of pigs carrying targeted mutations in genes has not been explored because, as yet, the technology to make such animals does not exist. The major factor preventing progress in this area is our inability to develop totipotent pig ES cells. The recent description of sheep cloning by the transfer of nuclei from adult somatic cells into enucleated oocytes gives hope for alternative approaches to the development of pigs carrying targeted mutations.

As far as hyperacute rejection is concerned, an obvious candidate for targeting is the α-1,3-galactosyltransferase gene, whose product is responsible for the addition of sugars to which xenoantibodies invariably react. Indeed α-1,3-galactosyltransferase knockout mice have been developed, and their cells show a reduced binding to and activation of human serum. How this mutation affects hyperacute xenograft rejection has yet to be reported. Although such mice develop and remain fairly normal, the possibility exists that deletion of this enzyme could have more severe consequences in other animals. Should this turn out to be the case, the use of conditional targeting constructs that allow deletion of the enzyme only shortly before transplantation could be extremely valuable.

Autologous Transplantation

One of the most exciting potentials of gene targeting in transplantation is its use to correct genetic defects in stem cells, which may then be retransplanted into the donor, a process to which we have already alluded. Gene therapy protocols analogous to this, but involving the insertion of DNA at random genomic loci, are currently being developed. Gene targeting in this context has many advantages over random insertion because it can be used to restore a defective gene to its normal working state and avoids the risk of insertional mutagenesis. It is encouraging that muscle precursor cells, with many of the properties expected of a muscle stem cell, have been successfully modified by gene targeting.

In common with many potential gene therapies, a potential drawback of this approach is that the recipient may develop immunity to the newly expressed gene product, not having acquired tolerance to it during development. That this may be a reality, yet not a problem, is illustrated by the finding that normal muscle cells transplanted into dystrophin-deficient mdx mice illicit dystrophin-specific antibodies that apparently are not deleterious to the muscle cells. In these experiments, marked cell-mediated immunity did not appear. If we restrict therapies to cells that do not have professional antigen-presenting potential, we would hope that immunity to the newly expressed gene would largely be avoided.

Genetically Modified Organism

Genetically modified (GM) barley grown by researchers in Germany.
The GM barley was investigated for its effects on soil quality.

Genetically modified organism (GMO) is the organism whose genome has been engineered in the laboratory in order to favour the expression of desired physiological traits or the generation of desired biological products. In conventional livestock production, crop farming, and even pet breeding, it has long been the practice to breed select individuals of a species in order to produce offspring that have desirable traits. In genetic modification, however, recombinant genetic technologies are employed to produce organisms whose genomes have been precisely altered at the molecular level, usually by the inclusion of genes from unrelated species of organisms that code for traits that would not be obtained easily through conventional selective breeding.

Genetically modified organisms (GMOs) are produced using scientific methods that include recombinant DNA technology and reproductive cloning. In reproductive cloning, a nucleus is extracted from a cell of the individual to be cloned and is inserted into the enucleated cytoplasm of a host egg (an enucleated egg is an egg cell that has had its own nucleus removed). The process results in the generation of an offspring that is genetically identical to the donor individual. The first animal produced by means of this cloning technique with a nucleus from an adult donor cell (as opposed to a donor embryo) was a sheep named Dolly, born in 1996. Since then a number of other animals, including pigs, horses, and dogs, have been generated by reproductive cloning technology. Recombinant DNA technology, on the other hand, involves the insertion of one or more individual genes from an organism of one species into the DNA (deoxyribonucleic acid) of another. Whole-genome replacement, involving the transplantation of one bacterial genome into the "cell body," or cytoplasm, of another microorganism, has been reported, although this technology is still limited to basic scientific applications.

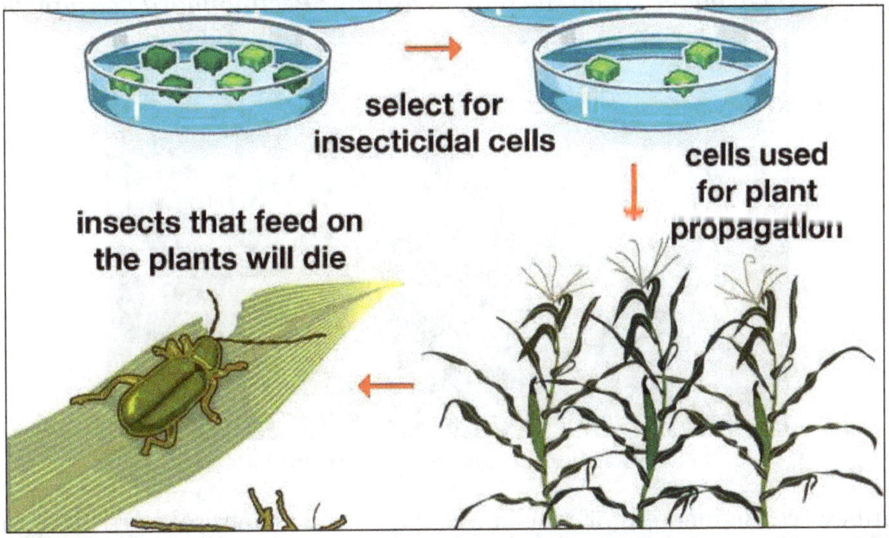

Genetically modified organisms are produced using scientific methods that include recombinant DNA technology.

GMOs produced through genetic technologies have become a part of everyday life, entering into society through agriculture, medicine, research, and environmental management. However, while GMOs have benefited human society in many ways, some disadvantages exist; therefore, the production of GMOs remains a highly controversial topic in many parts of the world.

GMOs in Agriculture

Genetically modified (GM) foods were first approved for human consumption in the United States in 1994, and by 2014–15 about 90 percent of the corn, cotton, and soybeans planted in the United States were GM. By the end of 2014, GM crops covered nearly 1.8 million square kilometres (695,000 square miles) of land in more than two dozen countries worldwide. The majority of GM crops were grown in the Americas.

Engineered crops can dramatically increase per area crop yields and, in some cases, reduce the use of chemical insecticides. For example, the application of wide-spectrum insecticides declined in many areas growing plants, such as potatoes, cotton, and corn, that were endowed with a gene from the bacterium *Bacillus thuringiensis*, which produces a natural insecticide called Bt toxin. Field studies conducted in India in which Bt cotton was compared with non-Bt cotton demonstrated a 30–80 percent increase in yield from the GM crop. This increase was attributed to marked improvement in the GM plants' ability to overcome bollworm infestation, which was otherwise common. Studies of Bt cotton production in Arizona, U.S., demonstrated only small gains in yield—about 5 percent—with an estimated cost reduction of $25–$65 (USD) per acre owing to decreased pesticide applications. In China, where farmers first gained access to Bt cotton in 1997, the GM crop was initially successful. Farmers who had planted Bt cotton reduced their pesticide use by 50–80 percent and increased their earnings by as much as 36 percent. By 2004, however, farmers who had been growing Bt cotton for several years found that the benefits of the crop eroded as populations of secondary insect pests, such as mirids, increased. Farmers once again were forced to spray broad-spectrum pesticides throughout the growing season, such that the average revenue for Bt growers was 8 percent lower than that of farmers who grew conventional cotton. Meanwhile, Bt resistance had also evolved in field populations of major cotton pests, including both the cotton bollworm (*Helicoverpa armigera*) and the pink bollworm (*Pectinophora gossypiella*).

Genetically engineered corn (maize).

Other GM plants were engineered for resistance to a specific chemical herbicide, rather than resistance to a natural predator or pest. Herbicide-resistant crops (HRC) have been available since the mid-1980s; these crops enable effective chemical control of weeds, since only the HRC plants can survive in fields treated with the corresponding herbicide. Many HRCs are resistant to glyphosate (Roundup), enabling liberal application of the chemical, which is highly effective against weeds. Such crops have been especially valuable for no-till farming, which helps prevent soil erosion. However, because HRCs encourage increased application of chemicals to the soil, rather than decreased application, they remain controversial with regard to their environmental impact. In addition, in order to reduce the risk of selecting for herbicide-resistant weeds, farmers must use multiple diverse weed-management strategies.

Another example of a GM crop is "golden" rice, which originally was intended for Asia and was genetically modified to produce almost 20 times the beta-carotene of previous varieties. Golden rice was created by modifying the rice genome to include a gene from the daffodil Narcissus pseudonarcissus that produces an enzyme known as phyotene synthase and a gene from the bacterium Erwinia uredovora that produces an enzyme called phyotene desaturase. The introduction of these genes enabled beta-carotene, which is converted to vitamin A in the human liver, to accumulate in the rice endosperm—the edible part of the rice plant—thereby increasing the amount of beta-carotene available for vitamin A synthesis in the body. In 2004 the same researchers who developed the original golden rice plant improved upon the model, generating golden rice 2, which showed a 23-fold increase in carotenoid production.

Another form of modified rice was generated to help combat iron deficiency, which impacts close to 30 percent of the world population. This GM crop was engineered by introducing into the rice genome a ferritin gene from the common bean, Phaseolus vulgaris that produces a protein capable of binding iron, as well as a gene from the fungus Aspergillus fumigatus that produces an enzyme capable of digesting compounds that increase iron bioavailability via digestion of phytate (an inhibitor of iron absorption). The iron-fortified GM rice was engineered to overexpress an existing rice gene that produces a cysteine-rich metallothioneinlike (metal-binding) protein that enhances iron absorption. A variety of other crops modified to endure the weather extremes common in other parts of the globe are also in production.

GMOs in Medicine and Research

GMOs have emerged as one of the mainstays of biomedical research since the 1980s. For example, GM animal models of human genetic diseases enabled researchers to test novel therapies and to explore the roles of candidate risk factors and modifiers of disease outcome. GM microbes, plants, and animals also revolutionized the production of complex pharmaceuticals by enabling the generation of safer and cheaper vaccines and therapeutics. Pharmaceutical products range from recombinant hepatitis B vaccine produced by GM baker's yeast to injectable insulin (for diabetics) produced in GM *Escherichia coli* bacteria and to factor VIII (for hemophiliacs) and tissue plasminogen activator (tPA, for heart attack or stroke patients), both of which are produced in GM mammalian cells grown in laboratory culture. Furthermore, GM plants that produce "edible vaccines" are under development. An edible vaccine is an antigenic protein that is produced in the consumable parts of a plant (e.g., fruit) and absorbed into the bloodstream when the parts are eaten. Once absorbed into the body, the protein stimulates the immune system to produce antibodies against the pathogen from which the antigen was derived. Such vaccines could offer a safe, inexpensive, and painless way to provide vaccines, particularly in less-developed regions of the world, where the limited availability of refrigeration and sterile needles has been problematic for some traditional vaccines. Novel DNA vaccines may be useful in the struggle to prevent diseases that have proved resistant to traditional vaccination approaches, including HIV/AIDS, tuberculosis, and cancer.

Genetic modification of insects has become an important area of research, especially in the struggle to prevent parasitic diseases. For example, GM mosquitoes have been developed that express a small protein called SM1, which blocks entry of the malaria parasite, *Plasmodium*, into the mosquito's gut. This results in the disruption of the parasite's life cycle and renders the mosquito malaria-resistant. Introduction of these GM mosquitoes into the wild could help reduce transmission of the malaria parasite. In another example, male Aedes aegypti mosquitoes engineered with a method known as the sterile insect technique transmit a gene to their offspring that causes the offspring to die before becoming sexually mature. In field trials in a Brazil suburb, A. aegypti populations declined by 95 percent following the sustained release of sterile GM males.

Finally, genetic modification of humans via gene therapy is becoming a treatment option for diseases ranging from rare metabolic disorders to cancer. Coupling stem cell technology with recombinant DNA methods allows stem cells derived from a patient to be modified in the laboratory to introduce a desired gene. For example, a normal beta-globin gene may be introduced into the DNA of bone marrow-derived hematopoietic stem cells from a patient with sickle cell anemia; introduction of these GM cells into the patient could cure the disease without the need for a matched donor.

Role of GMOs in Environmental Management

Another application of GMOs is in the management of environmental issues. For example, some bacteria can produce biodegradable plastics, and the transfer of that ability to microbes that can be easily grown in the laboratory may enable the wide-scale "greening" of the plastics industry. In the early 1990s, Zeneca, a British company, developed a microbially produced biodegradable plastic called Biopol (polyhydroxyalkanoate, or PHA). The plastic was made with the use of a GM bacterium, Ralstonia eutropha, to convert glucose and a variety of organic acids into a flexible polymer. GMOs endowed with the bacterially encoded ability to metabolize oil and heavy metals may provide efficient bioremediation strategies.

Socio Political Relevance of GMOs

While GMOs offer many potential benefits to society, the potential risks associated with them have fueled controversy, especially in the food industry. Many skeptics warn about the dangers that GM crops may pose to human health. For example, genetic manipulation may potentially alter the allergenic properties of crops. Whether some GM crops, such as golden rice, deliver on the promise of improved health benefits is also unclear. The release of GM mosquitoes and other GMOs into the environment also raised concerns. More-established risks were associated with the potential spread of engineered crop genes to native flora and the possible evolution of insecticide-resistant "superbugs."

From the late 1990s, the European Union (EU) addressed such concerns by implementing strict GMO labeling laws. In the early 2000s, all GM foods and GM animal feeds in the EU were required to be labeled if they consisted of or contained GM products in a proportion greater than 0.9 percent. By contrast, in the United States, foods containing GM ingredients did not require special labeling, though the issue was hotly debated at national and state levels. Many opponents of GM products focused their arguments on unknown risks to food safety. However, despite the concerns of some consumer and health groups, especially in Europe, numerous scientific panels, including the U.S. Food and Drug Administration, concluded that consumption of GM foods was safe, even in cases involving GM foods with genetic material from very distantly related organisms.

The strict regulations on GM products in the EU have been a source of tension in agricultural trade. In the late 1990s, the EU declared a moratorium on the import and use of GM crops. However, the ban—which led to trade disputes with other countries, particularly the United States, where GM foods had been accepted openly—was considered unjustified by the World Trade Organization. In consequence, the EU implemented regulatory changes that allowed for the import of certain GM crops. Within Europe, however, only one GM crop, a type of insect-resistant corn (maize), was cultivated. Some countries, including certain African states, had likewise rejected GM products. Still other countries, such as Canada, China, Argentina, and Australia, had open policies on GM foods.

The use of GMOs in medicine and research has produced a debate that is more philosophical in nature. For example, while genetic researchers believe they are working to cure disease and ameliorate suffering, many people worry that current gene therapy approaches may one day be applied to produce "designer" children or to lengthen the natural human life span. Similar to many other technologies, gene therapy and the production and application of GMOs can be used to address and resolve complicated scientific, medical, and environmental issues, but they must be used wisely.

References

- What-is-genetic-engineering, science-education-10806: jove.com, Retrieved 02, May 2020
- A-Cloning-and-Genetic-Engineering, A-Concepts-in-Biology-(OpenStax), Introductory-and-General-Biology: libretexts.org, Retrieved 13, January 2020
- Gene-therapy, science: britannica.com, Retrieved 26, March 2020
- Gene-targeting-techniques-and-applications-11150: lww.com, Retrieved 10, June 2020
- Genetically-modified-organism, science: britannica.com, Retrieved 19, April 2020

5

Human Genetics and Genomics

Human genetics is a subfield of genetics that is concerned with the study of the human genome and inheritance in human beings. The complete set of nucleic acid sequence encoded in the DNA of humans is referred to as the human genome. This chapter closely examines the key concepts of human genetics to provide an extensive understanding of the subject.

Human Genetics

Human Genetics is the study of inheritance as it occurs in human beings. Human genetics encompasses a variety of overlapping fields including: classical genetics, cytogenetics, molecular genetics, biochemical genetics, genomics, population genetics, developmental genetics, clinical genetics, and genetic counseling.

Genes can be the common factor of the qualities of most human-inherited traits. Study of human genetics can be useful as it can answer questions about human nature, understand the diseases and development of effective disease treatment, and understand genetics of human life. This article describes only basic features of human genetics; for the genetics of disorders please see: Medical genetics.

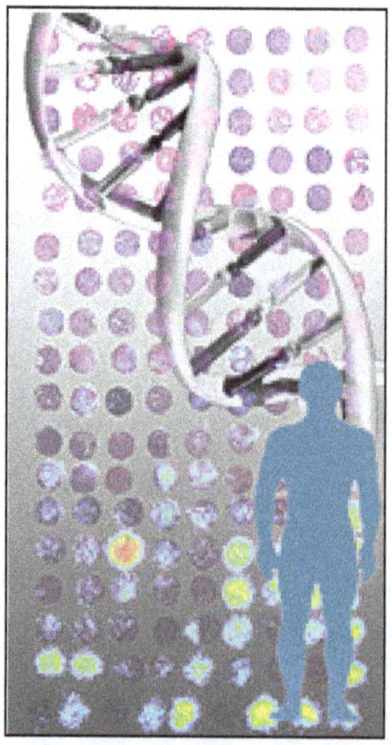

A small piece of human DNA.

Genetic Differences and Inheritance Patterns

Inheritance of traits for humans are based upon Gregor Mendel's model of inheritance. Mendel deduced that inheritance depends upon discrete units of inheritance, called factors or genes.

Autosomal Dominant Inheritance

Autosomal traits are associated with a single gene on an autosome (non-sex chromosome)—they are called "dominant" because a single copy—inherited from either parent—is enough to cause this trait to appear. This often means that one of the parents must also have the same trait, unless it has arisen due to an unlikely new mutation. Examples of autosomal dominant traits and disorders are Huntington's disease and achondroplasia.

Autosomal Recessive Inheritance

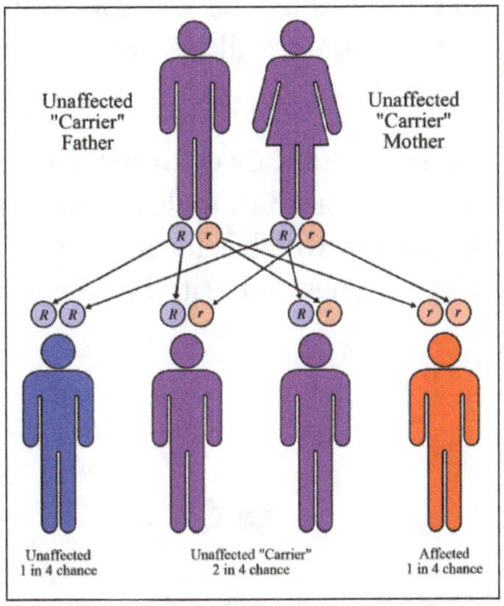

Autosomal recessive inheritance, a 25% chance.

Autosomal recessive traits is one pattern of inheritance for a trait, disease, or disorder to be passed on through families. For a recessive trait or disease to be displayed two copies of the trait or disorder needs to be presented. The trait or gene will be located on a non-sex chromosome. Because it takes two copies of a trait to display a trait, many people can unknowingly be carriers of a disease. From an evolutionary perspective, a recessive disease or trait can remain hidden for several generations before displaying the phenotype. Examples of autosomal recessive disorders are albinism, cystic fibrosis.

X-linked and Y-linked Inheritance

X-linked genes are found on the sex X chromosome. X-linked genes just like autosomal genes have both dominant and recessive types. Recessive X-linked disorders are rarely seen in females and usually only affect males. This is because males inherit their X chromosome and all X-linked genes will be inherited from the maternal side. Fathers only pass on their Y chromosome to their sons, so no X-linked traits will be inherited from father to son. Men cannot be carriers for recessive X linked traits, as they only have one X chromosome, so any X linked trait inherited from the mother will show up.

Females express X-linked disorders when they are homozygous for the disorder and become carriers when they are heterozygous. X-linked dominant inheritance will show the same phenotype as a heterozygote and homozygote. Just like X-linked inheritance, there will be a lack of male-to-male inheritance, which makes it distinguishable from autosomal traits. One example of an X-linked trait is Coffin–Lowry syndrome, which is caused by a mutation in ribosomal protein gene. This mutation results in skeletal, craniofacial abnormalities, mental retardation, and short stature.

X chromosomes in females undergo a process known as X inactivation. X inactivation is when one of the two X chromosomes in females is almost completely inactivated. It is important that this process occurs otherwise a woman would produce twice the amount of normal X chromosome proteins. The mechanism for X inactivation will occur during the embryonic stage. For people with disorders like trisomy X, where the genotype has three X chromosomes, X-inactivation will inactivate all X chromosomes until there is only one X chromosome active. Males with Klinefelter syndrome, who have an extra X chromosome, will also undergo X inactivation to have only one completely active X chromosome.

Y-linked inheritance occurs when a gene, trait, or disorder is transferred through the Y chromosome. Since Y chromosomes can only be found in males, Y linked traits are only passed on from father to son. The testis determining factor, which is located on the Y chromosome, determines the maleness of individuals. Besides the maleness inherited in the Y-chromosome there are no other found Y-linked characteristics.

Pedigrees

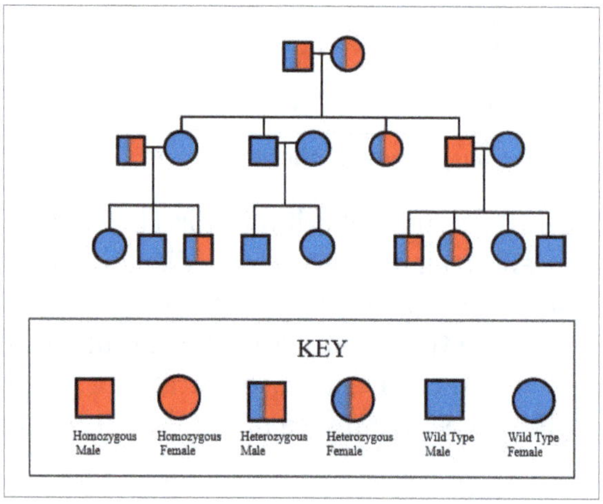

An example of a family pedigree displaying an autosomal recessive trait.

A pedigree is a diagram showing the ancestral relationships and transmission of genetic traits over several generations in a family. Square symbols are almost always used to represent males, whilst circles are used for females. Pedigrees are used to help detect many different genetic diseases. A pedigree can also be used to help determine the chances for a parent to produce an offspring with a specific trait.

Four different traits can be identified by pedigree chart analysis: autosomal dominant, autosomal recessive, x-linked, or y-linked. Partial penetrance can be shown and calculated form pedigrees. Penetrance is the percentage expressed frequency with which individuals of a given genotype manifest at least some degree of a specific mutant phenotype associated with a trait.

Inbreeding, or mating between closely related organisms, can clearly be seen on pedigree charts.

Pedigree charts of royal families often have a high degree of inbreeding, because it was customary and preferable for royalty to marry another member of royalty. Genetic counselors commonly use pedigrees to help couples determine if the parents will be able to produce healthy children.

Karyotype

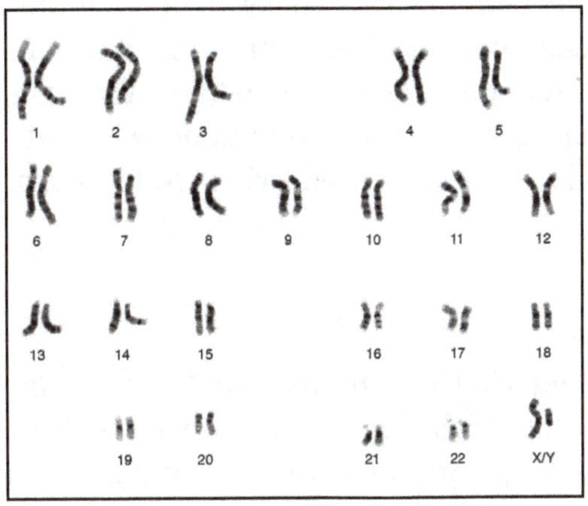

A karyotype of a human male, showing 46 chromosomes including XY sex chromosomes.

A karyotype is a very useful tool in cytogenetics. A karyotype is picture of all the chromosomes in the metaphase stage arranged according to length and centromere position. A karyotype can also be useful in clinical genetics, due to its ability to diagnose genetic disorders. On a normal karyotype, aneuploidy can be detected by clearly being able to observe any missing or extra chromosomes.

Giemsa banding, g-banding, of the karyotype can be used to detect deletions, insertions, duplications, inversions, and translocations. G-banding will stain the chromosomes with light and dark bands unique to each chromosome. A FISH, fluorescent in situ hybridization, can be used to observe deletions, insertions, and translocations. FISH uses fluorescent probes to bind to specific sequences of the chromosomes that will cause the chromosomes to fluoresce a unique color.

Genomics

Genomics refers to the field of genetics concerned with structural and functional studies of the genome. A genome is all the DNA contained within an organism or a cell including nuclear and mitochondrial DNA. The human genome is the total collection of genes in a human being contained in the human chromosome, composed of over three billion nucleotides. In April 2003, the Human Genome Project was able to sequence all the DNA in the human genome, and to discover that the human genome was composed of around 20,000 protein coding genes.

Medical Genetics

Medical genetics' is the branch of medicine that involves the diagnosis and management of hereditary disorders. Medical genetics is the application of genetics to medical care. It overlaps human genetics, for example, research on the causes and inheritance of genetic disorders would be considered within both human genetics and medical genetics, while the diagnosis, management, and counseling of individuals with genetic disorders would be considered part of medical genetics.

Population Genetics

Population genetics is the branch of evolutionary biology responsible for investigating processes that cause changes in allele and genotype frequencies in populations based upon Mendelian inheritance. Four different forces can influence the frequencies: natural selection, mutation, gene flow (migration), and genetic drift. A population can be defined as a group of interbreeding individuals and their offspring. For human genetics the populations will consist only of the human species. The Hardy-Weinberg principle is a widely used principle to determine allelic and genotype frequencies.

Mitochondrial DNA

In addition to nuclear DNA, humans (like almost all eukaryotes) have mitochondrial DNA. Mitochondria, the "power houses" of a cell, have their own DNA. Mitochondria are inherited from one's mother, and its DNA is frequently used to trace maternal lines of descent. Mitochondrial DNA is only 16kb in length and encodes for 62 genes.

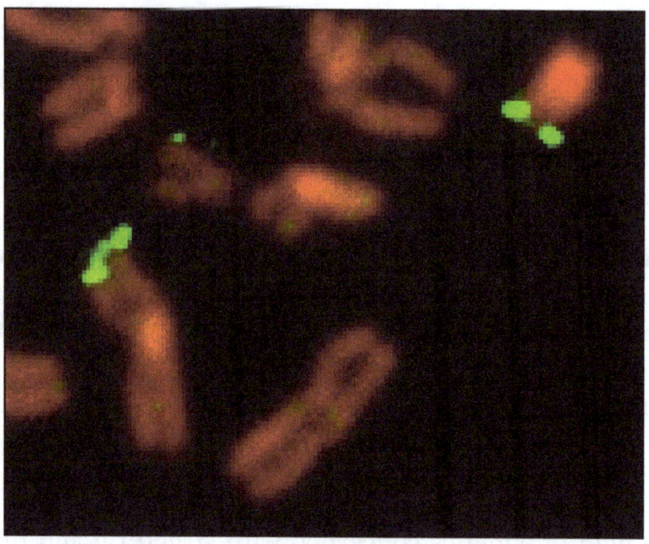

XY Chromosomes.

Genes and Sex

The XY sex-determination system is the sex-determination system found in humans, most other mammals, some insects (*Drosophila*), and some plants (*Ginkgo*). In this system, the sex of an individual is determined by a pair of sex chromosomes (gonosomes). Females have two of the same kind of sex chromosome (XX), and are called the homogametic sex. Males have two distinct sex chromosomes (XY), and are called the heterogametic sex.

X-linked Traits

Sex linkage is the phenotypic expression of an allele related to the chromosomal sex of the individual. This mode of inheritance is in contrast to the inheritance of traits on autosomal chromosomes, where both sexes have the same probability of inheritance. Since humans have many more genes on the X than the Y, there are many more X-linked traits than Y-linked traits. However, females carry two or more copies of the X chromosome, resulting in a potentially toxic dose of X-linked genes.

To correct this imbalance, mammalian females have evolved a unique mechanism of dosage compensation. In particular, by way of the process called X-chromosome inactivation (XCI), female mammals transcriptionally silence one of their two Xs in a complex and highly coordinated manner.

X-link dominant	X-link recessive	References
Alport syndrome	Absence of blood in urine	
Coffin–Lowry syndrome	No cranial malformations	
Colour vision	Colour blindness	
Normal clotting factor	Haemophilia A & B	
Strong muscle tissue	Duchenne Muscular Dystrophy	
fragile X syndrome	Normal X chromosome	
Aicardi syndrome	Absence of brain defects	
Absence of autoimmunity	IPEX syndrome	
Xg blood type	Absence of antigen	
Production of GAGs	Hunter syndrome	
Normal muscle strength	Becker's Muscular Dystrophy	
Unaffected body	Fabry's disease	
No progressive blindness	Choroideremia	
No kidney damage	Dent's disease	
Rett syndrome	No microcephaly	
Production of HGPRT	Lesch–Nyhan syndrome	
High levels of copper	Menkes disease	
Normal immune levels	Wiskott–Aldrich syndrome	
Focal dermal hypoplasia	Normal pigmented skin	
Normal pigment in eyes	Ocular albinism	
Vitamin D resistant rickets	Absorption of Vitamin D	
Synesthesia	Non colour perception	

Human Traits with Simple Inheritance Patterns

Dominant	Recessive	References
Low heart rate	High heart rate	
Widow's peak	straight hair line	
ocular hypertelorism	Hypotelorism	
Normal digestive muscle	POLIP syndrome	
Facial dimples *	No facial dimples	
Able to taste PTC	Unable to taste PTC	

Unattached (free) earlobe	Attached earlobe	
Clockwise hair direction (left to right)	Counter-Clockwise hair direction (right to left)	
Cleft chin	smooth chin	
No progressive nerve damage	Friedreich's ataxia	
Ability to roll tongue (Able to hold tongue in a U shape)	No ability to roll tongue	
extra finger or toe	Normal five fingers and toes	
Straight Thumb	Hitchhiker's Thumb	
Freckles	No freckles	
Wet-type earwax	Dry-type earwax	
Normal flat palm	Cenani Lenz syndactylism	
shortness in fingers	Normal finger length	
Webbed fingers	Normal separated fingers	
Roman nose	No prominent bridge	
Marfan's syndrome	Normal body proportions	
Huntington's disease	No nerve damage	
Normal mucus lining	Cystic fibrosis	
Photic sneeze reflex	No ACHOO reflex	
Forged chin	Receding chin	
White Forelock	Dark Forelock	
Ligamentous angustus	Ligamentous Laxity	
Ability to eat sugar	Galactosemia	
Total leukonychia and Bart pumphrey syndrome	partial leukonychia	
Absence of fish-like body odour	Trimethylaminuria	
Primary Hyperhidrosis	little sweating in hands	
Lactose persistence *	Lactose intolerance *	
Prominent chin (V-shaped)	less prominent chin (U-shaped)	
Acne prone	Clear complexion	
Normal height	Cartilage–hair hypoplasia	

Handicapping Conditions

Genetic Chromosomal

Effect	Source	References
Down syndrome	Additional 21st chromosome	
Cri du chat syndrome	Partial deletion of a chromosome in the B Group	
Klinefelter syndrome	One or more extra sex chromosome(s)	
Turner syndrome	Rearrangement of one or both X chromosomes, deletion of part of the second X chromosome, presence of part of a Y chromosome	

Human Evolutionary Genetics

Human evolutionary genetics studies how one human genome differs from another human genome, the evolutionary past that gave rise to it, and its current effects. Differences between genomes have anthropological, medical and forensic implications and applications. Genetic data can provide important insight into human evolution.

Origin of Apes

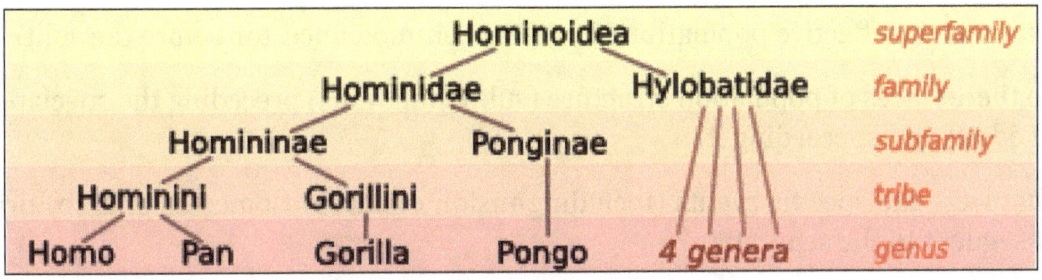

The taxonomic relationships of hominoids.

Biologists classify humans, along with only a few other species, as great apes (species in the family Hominidae). The living Hominidae include two distinct species of chimpanzee (the bonobo, *Pan paniscus*, and the common chimpanzee, *Pan troglodytes*), two species of gorilla (the western gorilla, *Gorilla gorilla*, and the eastern gorilla, *Gorilla graueri*), and two species of orangutan (the Bornean orangutan, *Pongo pygmaeus*, and the Sumatran orangutan, *Pongo abelii*). The great apes with the family Hylobatidae of gibbons form the superfamily Hominoidea of apes.

Apes, in turn, belong to the primate order (>400 species), along with the Old World monkeys, the New World monkeys, and others. Data from both mitochondrial DNA (mtDNA) and nuclear DNA (nDNA) indicate that primates belong to the group of Euarchontoglires, together with Rodentia, Lagomorpha, Dermoptera, and Scandentia. This is further supported by Alu-like short interspersed nuclear elements (SINEs) which have been found only in members of the Euarchontoglires.

Cladistics

A phylogenetic tree is usually derived from DNA or protein sequences from populations. Often, mitochondrial DNA or Y chromosome sequences are used to study ancient human demographics. These single-locus sources of DNA do not recombine and are almost always inherited from a single parent, with only one known exception in mtDNA. Individuals from closer geographic regions generally tend to be more similar than individuals from regions farther away. Distance on a phylogenetic tree can be used approximately to indicate:

1. Genetic distance. The genetic difference between humans and chimps is less than 2%, or twenty times larger than the variation among modern humans.
2. Temporal remoteness of the most recent common ancestor. The mitochondrial most recent common ancestor of modern humans lived roughly 200,000 years ago, latest common ancestors of humans and chimps between four and seven million years ago.

Speciation of Humans and the African Apes

The separation of humans from their closest relatives, the non-human apes (chimpanzees and gorillas), has been studied extensively for more than a century. Five major questions have been addressed:

- Which apes are our closest ancestors?
- When did the separations occur?
- What was the effective population size of the common ancestor before the split?
- Are there traces of population structure (subpopulations) preceding the speciation or partial admixture succeeding it?
- What were the specific events (including fusion of chromosomes 2a and 2b) prior to and subsequent to the separation?

General Observations

As discussed before, different parts of the genome show different sequence divergence between different hominoids. It has also been shown that the sequence divergence between DNA from humans and chimpanzees varies greatly. For example, the sequence divergence varies between 0% to 2.66% between non-coding, non-repetitive genomic regions of humans and chimpanzees. Additionally gene trees, generated by comparative analysis of DNA segments, do not always fit the species tree. Summing up:

- The sequence divergence varies significantly between humans, chimpanzees and gorillas.
- For most DNA sequences, humans and chimpanzees appear to be most closely related, but some point to a human-gorilla or chimpanzee-gorilla clade.
- The human genome has been sequenced, as well as the chimpanzee genome. Humans have 23 pairs of chromosomes, while chimpanzees, gorillas, and orangutans have 24. Human chromosome 2 is a fusion of two chromosomes 2a and 2b that remained separate in the other primates.

Divergence Times

The divergence time of humans from other apes is of great interest. One of the first molecular studies, published in 1967 measured immunological distances (IDs) between different primates. Basically the study measured the strength of immunological response that an antigen from one species (human albumin) induces in the immune system of another species (human, chimpanzee, gorilla and Old World monkeys). Closely related species should have similar antigens and therefore weaker immunological response to each other's antigens. The immunological response of a species to its own antigens (e.g. human to human) was set to be 1.

The ID between humans and gorillas was determined to be 1.09, that between humans and chimpanzees was determined as 1.14. However the distance to six different Old World monkeys was on average 2.46, indicating that the African apes are more closely related to humans than to monkeys.

The authors consider the divergence time between Old World monkeys and hominoids to be 30 million years ago (MYA), based on fossil data, and the immunological distance was considered to grow at a constant rate. They concluded that divergence time of humans and the African apes to be roughly ~5 MYA. That was a surprising result. Most scientists at that time thought that humans and great apes diverged much earlier (>15 MYA).

The gorilla was, in ID terms, closer to human than to chimpanzees; however, the difference was so slight that the trichotomy could not be resolved with certainty. Later studies based on molecular genetics were able to resolve the trichotomy: chimpanzees are phylogenetically closer to humans than to gorillas. However, some divergence times estimated later (using much more sophisticated methods in molecular genetics) do not substantially differ from the very first estimate in 1967, but a recent paper puts it at 11-14 MYA.

Divergence Times and Ancestral Effective Population Size

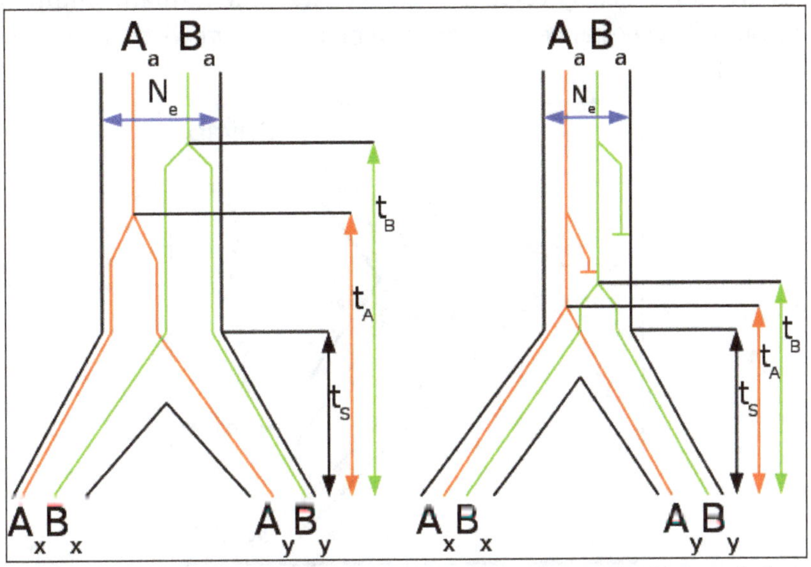

The sequences of the DNA segments diverge earlier than the species. A large effective population size in the ancestral population (left) preserves different variants of the DNA segments (=alleles) for a longer period of time. Therefore, on average, the gene divergence times (tA for DNA segment A; tB for DNA segment B) will deviate more from the time the species diverge (tS) compared to a small ancestral effective population size (right).

Current methods to determine divergence times use DNA sequence alignments and molecular clocks. Usually the molecular clock is calibrated assuming that the orangutan split from the African apes (including humans) 12-16 MYA. Some studies also include some old world monkeys and set the divergence time of them from hominoids to 25-30 MYA. Both calibration points are based on very little fossil data and have been criticized.

If these dates are revised, the divergence times estimated from molecular data will change as well. However, the relative divergence times are unlikely to change. Even if we can't tell absolute divergence times exactly, we can be pretty sure that the divergence time between chimpanzees and humans is about sixfold shorter than between chimpanzees (or humans) and monkeys.

One study (Takahata *et al.*, 1995) used 15 DNA sequences from different regions of the genome from human and chimpanzee and 7 DNA sequences from human, chimpanzee and gorilla. They determined that chimpanzees are more closely related to humans than gorillas. Using various statistical methods, they estimated the divergence time human-chimp to be 4.7 MYA and the divergence time between gorillas and humans (and chimps) to be 7.2 MYA.

Additionally they estimated the effective population size of the common ancestor of humans and chimpanzees to be ~100,000. This was somewhat surprising since the present day effective population size of humans is estimated to be only ~10,000. If true that means that the human lineage would have experienced an immense decrease of its effective population size (and thus genetic diversity) in its evolution.

Another study (Chen & Li, 2001) sequenced 53 non-repetitive, intergenic DNA segments from a human, a chimpanzee, a gorilla, and orangutan. When the DNA sequences were concatenated to a single long sequence, the generated neighbor-joining tree supported the *Homo-Pan* clade with 100% bootstrap (that is that humans and chimpanzees are the closest related species of the four). When three species are fairly closely related to each other (like human, chimpanzee and gorilla), the trees obtained from DNA sequence data may not be congruent with the tree that represents the speciation (species tree).

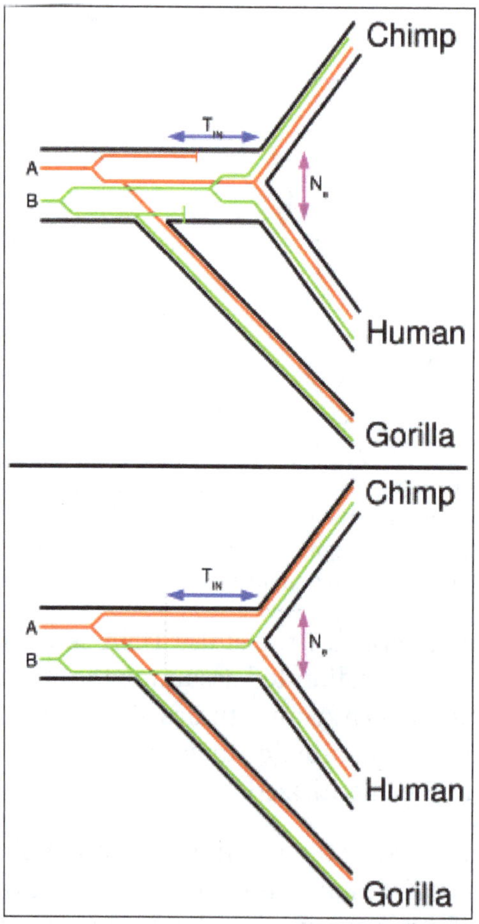

A and B are two different loci. In the upper figure they fit to the species tree. The DNA that is present in today's gorillas diverged earlier from the DNA that is present in today's humans and chimps. Thus both loci should be more similar between human and chimp than between gorilla and chimp or gorilla and human. In the lower graph, locus A has a more recent common ancestor in human and gorilla compared to the chimp sequence. Whereas chimp and gorilla have a more recent common ancestor for locus B. Here the gene trees are incongruent to the species tree.

The shorter internodal time span (T_{IN}) the more common are incongruent gene trees. The effective population size (N_e) of the internodal population determines how long genetic lineages are preserved in the population. A higher effective population size causes more incongruent gene trees. Therefore, if the internodal time span is known, the ancestral effective population size of the common ancestor of humans and chimpanzees can be calculated.

When each segment was analyzed individually, 31 supported the *Homo-Pan* clade, 10 supported the *Homo-Gorilla* clade, and 12 supported the *Pan-Gorilla* clade. Using the molecular clock the authors estimated that gorillas split up first 6.2-8.4 MYA and chimpanzees and humans split up 1.6-2.2 million years later (internodal time span) 4.6-6.2 MYA. The internodal time span is useful to estimate the ancestral effective population size of the common ancestor of humans and chimpanzees.

A parsimonious analysis revealed that 24 loci supported the *Homo-Pan* clade, 7 supported the *Homo-Gorilla* clade, 2 supported the *Pan-Gorilla* clade and 20 gave no resolution. Additionally they took 35 protein coding loci from databases. Of these 12 supported the *Homo-Pan* clade, 3 the *Homo-Gorilla* clade, 4 the *Pan-Gorilla* clade and 16 gave no resolution. Therefore, only ~70% of the 52 loci that gave a resolution (33 intergenic, 19 protein coding) support the 'correct' species tree. From the fraction of loci which did not support the species tree and the internodal time span they estimated previously, the effective population of the common ancestor of humans and chimpanzees was estimated to be ~52 000 to 96 000. This value is not as high as that from the first study (Takahata), but still much higher than present day effective population size of humans.

A third study (Yang, 2002) used the same dataset that Chen and Li used but estimated the ancestral effective population of 'only' ~12,000 to 21,000, using a different statistical method.

Genetic Differences Between Humans and Other Great Apes

The alignable sequences within genomes of humans and chimpanzees differ by about 35 million single-nucleotide substitutions. Additionally about 3% of the complete genomes differ by deletions, insertions and duplications.

Since mutation rate is relatively constant, roughly one half of these changes occurred in the human lineage. Only a very tiny fraction of those fixed differences gave rise to the different phenotypes of humans and chimpanzees and finding those is a great challenge. The vast majority of the differences are neutral and do not affect the phenotype.

Molecular evolution may act in different ways, through protein evolution, gene loss, differential gene regulation and RNA evolution. All are thought to have played some part in human evolution.

Gene Loss

Many different mutations can inactivate a gene, but few will change its function in a specific way. Inactivation mutations will therefore be readily available for selection to act on. Gene loss could thus be a common mechanism of evolutionary adaptation (the "less-is-more" hypothesis).

80 genes were lost in the human lineage after separation from the last common ancestor with the

chimpanzee. 36 of those were for olfactory receptors. Genes involved in chemoreception and immune response are overrepresented. Another study estimated that 86 genes had been lost.

Hair Keratin Gene Krthap1

A gene for type I hair keratin was lost in the human lineage. Keratins are a major component of hairs. Humans still have nine functional type I hair keratin genes, but the loss of that particular gene may have caused the thinning of human body hair. The gene loss occurred relatively recently in human evolution—less than 240,000 years ago.

Myosin Gene MYH16

Stedman *et al.* (2004) stated that the loss of the sarcomeric myosin gene MYH16 in the human lineage led to smaller masticatory muscles. They estimated that the mutation that led to the inactivation (a two base pair deletion) occurred 2.4 million years ago, predating the appearance of *Homo ergaster/erectus* in Africa. The period that followed was marked by a strong increase in cranial capacity, promoting speculation that the loss of the gene may have removed an evolutionary constraint on brain size in the genus *Homo*.

Another estimate for the loss of the MYH16 gene is 5.3 million years ago, long before *Homo* appeared.

Other

- CASPASE12, a cysteinyl aspartate proteinase. The loss of this gene is speculated to have reduced the lethality of bacterial infection in humans.

Gene Addition

Segmental duplications (SDs or LCRs) have had roles in creating new primate genes and shaping human genetic variation.

Human-specific DNA Insertions

When the human genome was compared to the genomes of five comparison primate species, including the chimpanzee, gorilla, orangutan, gibbon, and macaque, it was found that there are approximately 20,000 human-specific insertions believed to be regulatory. While most insertions appear to be fitness neutral, a small amount have been identified in positively selected genes showing associations to neural phenotypes and some relating to dental and sensory perception-related phenotypes. These findings hint at the seemingly important role of human-specific insertions in the recent evolution of humans.

Selection Pressures

Human accelerated regions are areas of the genome that differ between humans and chimpanzees to a greater extent than can be explained by genetic drift over the time since the two species shared a common ancestor. These regions show signs of being subject to natural selection, leading to the evolution of distinctly human traits. Two examples are HAR1F, which is believed to be related to brain development and HAR2 (a.k.a. HACNS1) that may have played a role in the development of the opposable thumb.

It has also been hypothesized that much of the difference between humans and chimpanzees is attributable to the regulation of gene expression rather than differences in the genes themselves. Analyses of conserved non-coding sequences, which often contain functional and thus positively selected regulatory regions, address this possibility.

Sequence Divergence Between Humans and Apes

When the draft sequence of the common chimpanzee (*Pan troglodytes*) genome was published in the summer 2005, 2400 million bases (of ~3160 million bases) were sequenced and assembled well enough to be compared to the human genome. 1.23% of this sequenced differed by single-base substitutions. Of this, 1.06% or less was thought to represent fixed differences between the species, with the rest being variant sites in humans or chimpanzees. Another type of difference, called indels (insertions/deletions) accounted for many fewer differences (15% as many), but contributed ~1.5% of unique sequence to each genome, since each insertion or deletion can involve anywhere from one base to millions of bases.

A companion paper examined segmental duplications in the two genomes, whose insertion and deletion into the genome account for much of the indel sequence. They found that a total of 2.7% of euchromatic sequence had been differentially duplicated in one or the other lineage.

Percentage sequence divergence between humans and other hominids			
Locus	**Human-Chimp**	**Human-Gorilla**	**Human-Orangutan**
Alu elements	2	-	-
Non-coding (Chr. Y)	1.68 ± 0.19	2.33 ± 0.2	5.63 ± 0.35
Pseudogenes (autosomal)	1.64 ± 0.10	1.87 ± 0.11	-
Pseudogenes (Chr. X)	1.47 ± 0.17	-	-
Noncoding (autosomal)	1.24 ± 0.07	1.62 ± 0.08	3.08 ± 0.11
Genes (K_s)	1.11	1.48	2.98
Introns	0.93 ± 0.08	1.23 ± 0.09	-
Xq13.3	0.92 ± 0.10	1.42 ± 0.12	3.00 ± 0.18
Subtotal for X chromosome	1.16 ± 0.07	1.47 ± 0.08	-
Genes (K_a)	0.8	0.93	1.96

The sequence divergence has generally the following pattern: Human-Chimp < Human-Gorilla << Human-Orangutan, highlighting the close kinship between humans and the African apes. Alu elements diverge quickly due to their high frequency of CpG dinucleotides which mutate roughly 10 times more often than the average nucleotide in the genome. The mutation rate is higher in the male germ line, therefore the divergence in the Y chromosome—which is inherited solely from the father—is higher than in autosomes. The X chromosome is inherited twice as often through the female germ line as through the male germ line and therefore shows slightly lower sequence divergence. The sequence divergence of the Xq13.3 region is surprisingly low between humans and chimpanzees.

Mutations altering the amino acid sequence of proteins (K_a) are the least common. In fact ~29% of all orthologous proteins are identical between human and chimpanzee. The typical protein differs by only two amino acids. The measures of sequence divergence shown in the table only take the substitutional differences, for example from an A (adenine) to a G (guanine), into account. DNA sequences may however also differ by insertions and deletions (indels) of bases. These are usually stripped from the alignments before the calculation of sequence divergence is performed.

Genetic Differences between Modern Humans and Neanderthals

An international group of scientists completed a draft sequence of the Neanderthal genome in May 2010. The results indicate some breeding between modern humans (*Homo sapiens*) and Neanderthals (*Homo neanderthalensis*), as the genomes of non-African humans have 1-4% more in common with Neanderthals than do the genomes of subsaharan Africans. Neanderthals and most modern humans share a lactose-intolerant variant of the lactase gene that encodes an enzyme that is unable to break down lactose in milk after weaning. Modern humans and Neanderthals also share the FOXP2 gene variant associated with brain development and with speech in modern humans, indicating that Neanderthals may have been able to speak. Chimps have two amino acid differences in FOXP2 compared with human and Neanderthal FOXP2.

Modern Humans

Molecular biologists starting with Wesley Brown on mtDNA and Allan Wilson on mtDNA have produced observations relevant to human evolution.

Age of the Common Ancestor

By estimating the rate at which mutations occur in mtDNA, the age of the common ancestral mtDNA type can be estimated: "the common ancestral mtDNA (type a) links mtDNA types that have diverged by an average of nearly 0.57%. Assuming a rate of 2%-4% per million years, this implies that the common ancestor of all surviving mtDNA types existed 140,000-290,000 years ago."

Most, but not all (fr. Multiregional origin of modern humans), scientists in the relevant fields consider this observation robust. This common direct ancestor in the line of mother to daughter (or mitochondrial most recent common ancestor (mtMRCA)) of all extant humans has become known as Mitochondrial Eve. (Mitochondria are inherited from the mother only.) The observation that the mtMRCA is the direct matrilineal ancestor of all living humans does not mean either that she was the first anatomically modern woman, nor that no other women lived concurrently with her, nor even excluding the existence of other women being ancestors of today's people. Other women would have lived at the same time and passed nuclear genes down to living humans, but their mitochondrial lineages were lost over time. This could be due to events such as producing only male children.

African Origin for Modern Humans

There is evidence that modern human mtDNA has an African origin: "We infer from the tree of minimum length... that Africa is a likely source of the human mitochondrial gene pool. This inference comes from the observation that one of the two primary branches leads exclusively to African mtDNAs... while the second primary branch also leads to African mtDNAs... By postulating that the common ancestral mtDNA... was African, we minimize the number of intercontinental migrations needed to account for the geographic distribution of mtDNA types."

The broad study of African genetic diversity headed by Sarah Tishkoff found the San people to express the greatest genetic diversity among the 113 distinct populations sampled, making them one of 14 "ancestral population clusters". The research also located the origin of modern human migration in south-western Africa, near the coastal border of Namibia and Angola.

Y Chromosome Lineages

The Y chromosome is much larger than mtDNA, and is relatively homogeneous; therefore it has taken much longer to find distinct lineages and to analyse them. Conversely, because the Y chromosome is so large by comparison, it holds more genetic information. Y chromosome studies show similar findings to those made with mtDNA. The estimate for the age of the ancestral Y chromosome for all extant Y chromosomes is given at about 70,000 years ago and is also placed in Africa; the individual who contributed this Y chromosomal heritage is sometimes referred to as Y chromosome Adam. The difference in dates between Y chromosome Adam and mitochondrial Eve is usually attributed to a higher extinction rate for Y chromosomes due to greater differential reproductive success between individual men, which means that a small number of very successful men may produce many children, while a larger number of less successful men will produce far fewer children.

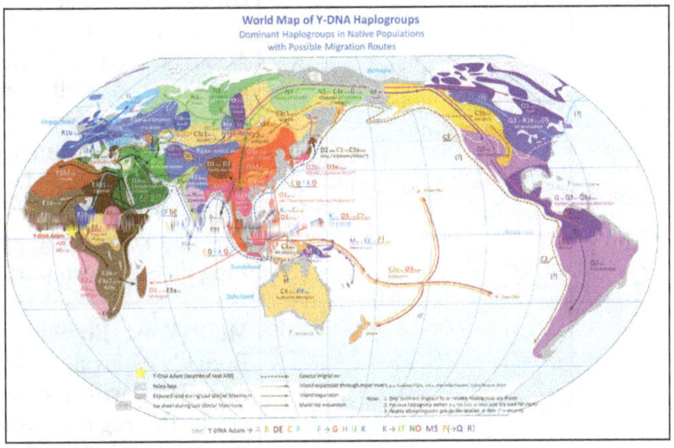

Dominant Y-chromosome haplogroups in pre-colonial world populations, with possible migrations routes according to the Coastal Migration Model.

At a recent conference at the American Society of Human Genetics, there was a study presented by Melissa Wilson Sayres from the University of California Berkeley which called into question the amount to which people had more female ancestors than male. While it didn't deny that men have historically produced at a higher rate than females, a polygamous mating system doesn't explain how the y-chromosome is so uniform. "Other people have suggested it's as few as four females for one male. We find that it's probably skewed, but it's more like four females to three males" said Sayres.

Ancestral Components

Modern DNA sequencing has identified various ancestral components in contemporary populations. A number of these genetic elements have West Eurasian origins. They include the following ancestral components, with their geographical hubs and main associated populations:

#	West Eurasian component	Geographical hub	Peak population	Notes
1	Ancestral North Indian	North India, Pakistan	North Indians, Pakistanis	Main West Eurasian component in the Indian subcontinent. Peaks among Indo-European-speaking caste populations in the northern areas, but also found at significant frequencies among some Dravidian-speaking caste groups. Associated with either the arrival of Indo-European speakers from West Asia or Central Asia between 3,000 and 4,000 years before present, or with the spread of agriculture and West Asian crops beginning around 8,000-9,000 ybp, or with migrations from West Asia in the pre-agricultural period. Contrasted with the indigenous Ancestral South Indian component, which peaks among the Onge Andamanese inhabiting the Andaman Islands.
2	Arabian	Arabian peninsula	Yemenis, Saudis, Qataris, Bedouins	Main West Eurasian component in the Gulf region. Most closely associated with local Arabic, Semitic-speaking populations. Also found at significant frequencies in parts of the Levant, Egypt and Libya.
3	Coptic	Nile Valley	Copts, Beja, Afro-Asiatic Ethiopians, Sudanese Arabs, Nubians	Main West Eurasian component in Northeast Africa. Roughly equivalent with the Ethio-Somali component. Peaks among Egyptian Copts in Sudan. Also found at high frequencies among other Afro-Asiatic (Hamito-Semitic) speakers in Ethiopia and Sudan, as well as among many Nubians. Associated with Ancient Egyptian ancestry, without the later Arabian influence present among modern Egyptians. Contrasted with the indigenous Nilo-Saharan component, which peaks among Nilo-Saharan- and Kordofanian-speaking populations inhabiting the southern part of the Nile Valley.
4	Ethio-Somali	Horn of Africa	Somalis, Afars, Amhara, Oromos, Tigrinya	Main West Eurasian component in the Horn. Roughly equivalent with the Coptic component. Associated with the arrival of Afro-Asiatic speakers in the region during antiquity. Peaks among Cushitic- and Ethiopian Semitic-speaking populations in the northern areas. Diverged from the Maghrebi component around 23,000 ybp, and from the Arabian component about 25,000 ybp. Contrasted with the indigenous Omotic component, which peaks among the Omotic-speaking Ari ironworkers inhabiting southern Ethiopia.
5	European	Europe	Europeans	Main West Eurasian component in Europe. Also found at significant frequencies in adjacent geographical areas outside of the continent, in Anatolia, the Caucasus, the Iranian plateau, and parts of the Levant.

#	West Eurasian component	Geographical hub	Peak population	Notes
6	Levantine	Near East, Caucasus	Druze, Lebanese, Cypriots, Syrians, Jordanians, Palestinians, Armenians, Georgians, Sephardic Jews, Ashkenazi Jews, Iranians, Turks, Sardinians, Adygei	Main West Eurasian component in the Near East and Caucasus. Peaks among Druze populations in the Levant. Found amongst local Afro-Asiatic, Indo-European, Caucasus and Turkish speakers alike. Diverged from the European component around 9,100-15,900 ybp, and from the Arabian component about 15,500-23,700 ypb. Also found at significant frequencies in Southern Europe as well as parts of the Arabian peninsula.
7	Maghrebi	Northwest Africa	Berbers, Maghrebis, Sahrawis, Tuareg	Main West Eurasian component in the Maghreb. Peaks among the Berber (non-Arabized) populations in the region. Diverged from the Ethio-Somali/Coptic, Arabian, Levantine and European components prior to the Holocene.

Human Genome

The human genome is the complete set of nucleic acid sequence for humans (*Homo sapiens*), encoded as DNA within the 23 chromosome pairs in cell nuclei and in a small DNA molecule found within individual mitochondria. Human genomes include both protein-coding DNA genes and noncoding DNA. Haploid human genomes, which are contained in germ cells (the egg and sperm gamete cells created in the meiosis phase of sexual reproduction before fertilization creates a zygote) consist of three billion DNA base pairs, while diploid genomes (found in somatic cells) have twice the DNA content. While there are significant differences among the genomes of human individuals (on the order of 0.1%), these are considerably smaller than the differences between humans and their closest living relatives, the chimpanzees (approximately 4%) and bonobos.

The Human Genome Project produced the first complete sequences of individual human genomes, with the first draft sequence and initial analysis being published on February 12, 2001. The human genome was the first of all vertebrates to be completely sequenced. As of 2012, thousands of human genomes have been completely sequenced, and many more have been mapped at lower levels of resolution. The resulting data are used worldwide in biomedical science, anthropology, forensics and other branches of science. There is a widely held expectation that genomic studies will lead to advances in the diagnosis and treatment of diseases, and to new insights in many fields of biology, including human evolution.

Although the sequence of the human genome has been (almost) completely determined by DNA sequencing, it is not yet fully understood. Most (though probably not all) genes have been identified by a combination of high throughput experimental and bioinformatics approaches, yet much work still needs to be done to further elucidate the biological functions of their protein and RNA products. Recent results suggest that most of thc vast quantities of noncoding DNA within the genome have associated biochemical activities, including regulation of gene expression, organization of chromosome architecture, and signals controlling epigenetic inheritance.

There are an estimated 20,000-25,000 human protein-coding genes. The estimate of the number of human genes has been repeatedly revised down from initial predictions of 100,000 or more as genome sequence quality and gene finding methods have improved, and could continue to drop further. Protein-coding sequences account for only a very small fraction of the genome (approximately 1.5%), and the rest is associated with non-coding RNA molecules, regulatory DNA sequences, LINEs, SINEs, introns, and sequences for which as yet no function has been determined.

In June 2016, scientists formally announced HGP-Write, a plan to synthesize the human genome.

Molecular Organization and Gene Content

The total length of the human genome is over 3 billion base pairs. The genome is organized into 22 paired chromosomes, plus the X chromosome (one in males, two in females) and, in males only, one Y chromosome. These are all large linear DNA molecules contained within the cell nucleus. The genome also includes the mitochondrial DNA, a comparatively small circular molecule present in each mitochondrion. Basic information about these molecules and their gene content, based on a reference genome that does not represent the sequence of any specific individual, are provided in the following table. (Data source: Ensembl genome browser release 68, July 2012).

Chromosome	Length (mm)	Base pairs	Variations	Confirmed proteins	Putative proteins	Pseudogenes	miRNA	rRNA	snRNA	snoRNA	Misc ncRNA	Links	Centromere position (Mbp)	Cumulative (%)
1	85	249,250,621	4,401,091	2,012	31	1,130	134	66	221	145	106	EBI	125.0	7.9
2	83	243,199,373	4,607,702	1,203	50	948	115	40	161	117	93	EBI	93.3	16.2
3	67	198,022,430	3,894,345	1,040	25	719	99	29	138	87	77	EBI	91.0	23.0
4	65	191,154,276	3,673,892	718	39	698	92	24	120	56	71	EBI	50.4	29.6
5	62	180,915,260	3,436,667	849	24	676	83	25	106	61	68	EBI	48.4	35.8
6	58	171,115,067	3,360,890	1,002	39	731	81	26	111	73	67	EBI	61.0	41.6
7	54	159,138,663	3,045,992	866	34	803	90	24	90	76	70	EBI	59.9	47.1
8	50	146,364,022	2,890,692	659	39	568	80	28	86	52	42	EBI	45.6	52.0
9	48	141,213,431	2,581,827	785	15	714	69	19	66	51	55	EBI	49.0	56.3
10	46	135,534,747	2,609,802	745	18	500	64	32	87	56	56	EBI	40.2	60.9
11	46	135,006,516	2,607,254	1,258	48	775	63	24	74	76	53	EBI	53.7	65.4
12	45	133,851,895	2,482,194	1,003	47	582	72	27	106	62	69	EBI	35.8	70.0
13	39	115,169,878	1,814,242	318	8	323	42	16	45	34	36	EBI	17.9	73.4
14	36	107,349,540	1,712,799	601	50	472	92	10	65	97	46	EBI	17.6	76.4
15	35	102,531,392	1,577,346	562	43	473	78	13	63	136	39	EBI	19.0	79.3
16	31	90,354,753	1,747,136	805	65	429	52	32	53	58	34	EBI	36.6	82.0
17	28	81,195,210	1,491,841	1,158	44	300	61	15	80	71	46	EBI	24.0	84.8
18	27	78,077,248	1,448,602	268	20	59	32	13	51	36	25	EBI	17.2	87.4
19	20	59,128,983	1,171,356	1,399	26	181	110	13	29	31	15	EBI	26.5	89.3
20	21	63,025,520	1,206,753	533	13	213	57	15	46	37	34	EBI	27.5	91.4
21	16	48,129,895	787,784	225	8	150	16	5	21	19	8	EBI	13.2	92.6
22	17	51,304,566	745,778	431	21	308	31	5	23	23	23	EBI	14.7	93.8
X	53	155,270,560	2,174,952	815	23	780	128	22	85	64	52	EBI	60.6	99.1

Y	20	59,373,566	286,812	45	8	327	15	7	17	3	2	EBI	12.5	**100.0**
mtDNA	0.0054	16,569	929	13	0	0	0	2	0	0	22	EBI	N/A	**100.0**
total		**3,095,693,981**		19,313	738	12,859								

Table 1 (above) summarizes the physical organization and gene content of the human reference genome, with links to the original analysis, as published in the Ensembl database at the European Bioinformatics Institute (EBI) and Wellcome Trust Sanger Institute. Chromosome lengths were estimated by multiplying the number of base pairs by 0.34 nanometers, the distance between base pairs in the DNA double helix. The number of proteins is based on the number of initial precursor mRNA transcripts, and does not include products of alternative pre-mRNA splicing, or modifications to protein structure that occur after translation.

The number of variations is a summary of unique DNA sequence changes that have been identified within the sequences analyzed by Ensembl as of July, 2012; that number is expected to increase as further personal genomes are sequenced and examined. In addition to the gene content shown in this table, a large number of non-expressed functional sequences have been identified throughout the human genome. Links open windows to the reference chromosome sequence in the EBI genome browser. The table also describes prevalence of genes encoding structural RNAs in the genome.

MicroRNA, or miRNA, functions as a post-transcriptional regulator of gene expression. Ribosomal RNA, or rRNA, makes up the RNA portion of the ribosome and is critical in the synthesis of proteins. Small nuclear RNA, or snRNA, is found in the nucleus of the cell. Its primary function is in the processing of pre-mRNA molecules and also in the regulation of transcription factors. Small nucleolar RNA, or SnoRNA, primarily functions in guiding chemical modifications to other RNA molecules.

Completeness of the Human Genome Sequence

Although the human genome has been completely sequenced for all practical purposes, there are still hundreds of gaps in the sequence. A recent study noted more than 160 euchromatic gaps of which 50 gaps were closed. However, there are still numerous gaps in the heterochromatic parts of the genome which is much harder to sequence due to numerous repeats and other intractable sequence features.

Coding vs. Noncoding DNA

The content of the human genome is commonly divided into coding and noncoding DNA sequences. Coding DNA is defined as those sequences that can be transcribed into mRNA and translated into proteins during the human life cycle; these sequences occupy only a small fraction of the genome (<2%). Noncoding DNA is made up of all of those sequences (ca. 98% of the genome) that are not used to encode proteins.

Some noncoding DNA contains genes for RNA molecules with important biological functions (noncoding RNA, for example ribosomal RNA and transfer RNA). The exploration of the function and evolutionary origin of noncoding DNA is an important goal of contemporary genome research, including the ENCODE (Encyclopedia of DNA Elements) project, which aims to survey the entire human genome, using a variety of experimental tools whose results are indicative of molecular activity.

Because non-coding DNA greatly outnumbers coding DNA, the concept of the sequenced genome has become a more focused analytical concept than the classical concept of the DNA-coding gene.

Mutation Rate of Human Genome

Mutation rate of human genome is a very important factor in calculating evolutionary time points. Researchers calculated the number of genetic variations between human and apes. Dividing that number by age of fossil of most recent common ancestor of humans and ape, researchers calculated the mutation rate. Recent studies using next generation sequencing technologies concluded a slow mutation rate which doesn't add up with human migration pattern time points and suggesting a new evolutionary time scale. 100,000 year old human fossils found in Israel have served to compound this new found uncertainty of the human migration timeline.

Coding Sequences (Protein-coding Genes)

Protein-coding sequences represent the most widely studied and best understood component of the human genome. These sequences ultimately lead to the production of all human proteins, although several biological processes (e.g. DNA rearrangements and alternative pre-mRNA splicing) can lead to the production of many more unique proteins than the number of protein-coding genes.

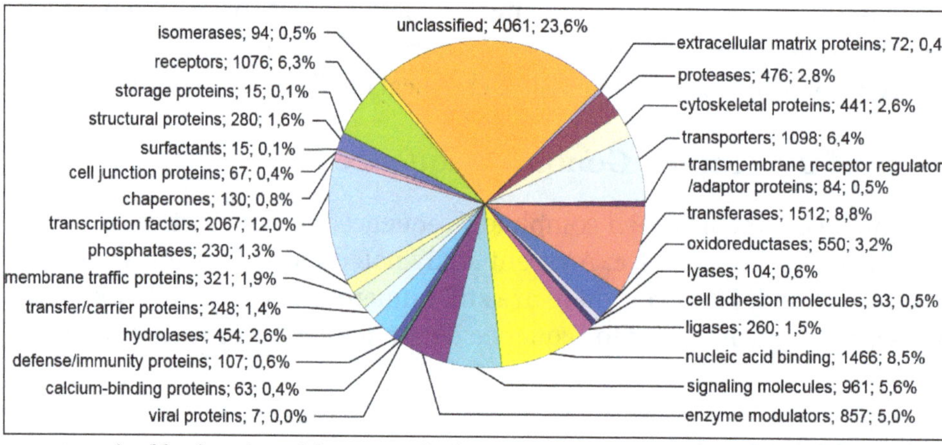

Human genes categorized by function of the transcribed proteins, given both as number of encoding genes and percentage of all genes.

The complete modular protein-coding capacity of the genome is contained within the exome, and consists of DNA sequences encoded by exons that can be translated into proteins. Because of its biological importance, and the fact that it constitutes less than 2% of the genome, sequencing of the exome was the first major milepost of the Human Genome Project.

Number of protein-coding genes. About 20,000 human proteins have been annotated in databases such as Uniprot. Historically, estimates for the number of protein genes have varied widely, ranging up to 2,000,000 in the late 1960s, but several researchers pointed out in the early 1970s that the estimated mutational load from deleterious mutations placed an upper limit of approximately 40,000 for the total number of functional loci (this includes protein-coding and functional non-coding genes).

The number of human protein-coding genes is not significantly larger than that of many less complex organisms, such as the roundworm and the fruit fly. This difference may result from the extensive use of alternative pre-mRNA splicing in humans, which provides the ability to build a very large number of modular proteins through the selective incorporation of exons.

Protein-coding capacity per chromosome. Protein-coding genes are distributed unevenly across the chromosomes, ranging from a few dozen to more than 2000, with an especially high gene density within chromosomes 19, 11, and 1 (Table 1). Each chromosome contains various gene-rich and gene-poor regions, which may be correlated with chromosome bands and GC-content. The significance of these nonrandom patterns of gene density is not well understood.

Size of protein-coding genes. The size of protein-coding genes within the human genome shows enormous variability (Table 2). For example, the gene for histone H1a (HIST1HIA) is relatively small and simple, lacking introns and encoding mRNA sequences of 781 nt and a 215 amino acid protein (648 nt open reading frame). Dystrophin (DMD) is the largest protein-coding gene in the human reference genome, spanning a total of 2.2 MB, while Titin (TTN) has the longest coding sequence (114,414 bp), the largest number of exons (363), and the longest single exon (17,106 bp). Over the whole genome, the median size of an exon is 122 bp (mean = 145 bp), the median number of exons is 7 (mean = 8.8), and the median coding sequence encodes 367 amino acids (mean = 447 amino acids; Table 21 in).

Protein	Chrom	Gene	Length	Exons	Exon length	Intron length	Alt splicing
Breast cancer type 2 susceptibility protein	13	BRCA2	83,736	27	11,386	72,350	yes
Cystic fibrosis transmembrane conductance regulator	7	CFTR	202,881	27	4,440	198,441	yes
Cytochrome b	MT	MTCYB	1,140	1	1,140	0	no
Dystrophin	X	DMD	2,220,381	79	10,500	2,209,881	yes
Glyceraldehyde-3-phosphate dehydrogenase	12	GAPDH	4,444	9	1,425	3,019	yes
Hemoglobin beta subunit	11	HBB	1,605	3	626	979	no
Histone H1A	6	HIST1H1A	781	1	781	0	no
Titin	2	TTN	281,434	364	104,301	177,133	yes

Table: Examples of human protein-coding genes. Chrom, chromosome. Alt splicing, alternative pre-mRNA splicing.

Noncoding DNA (ncDNA)

Noncoding DNA is defined as all of the DNA sequences within a genome that are not found within protein-coding exons, and so are never represented within the amino acid sequence of expressed proteins. By this definition, more than 98% of the human genomes is composed of ncDNA.

Numerous classes of noncoding DNA have been identified, including genes for noncoding RNA (e.g. tRNA and rRNA), pseudogenes, introns, untranslated regions of mRNA, regulatory DNA sequences, repetitive DNA sequences, and sequences related to mobile genetic elements.

Numerous sequences that are included within genes are also defined as noncoding DNA. These include genes for noncoding RNA (e.g. tRNA, rRNA), and untranslated components of protein-coding genes (e.g. introns, and 5' and 3' untranslated regions of mRNA).

Protein-coding sequences (specifically, coding exons) constitute less than 1.5% of the human genome. In addition, about 26% of the human genome is introns. Aside from genes (exons and introns) and known regulatory sequences (8–20%), the human genome contains regions of noncoding DNA. The exact amount of noncoding DNA that plays a role in cell physiology has been hotly debated. Recent analysis by the ENCODE project indicates that 80% of the entire human genome is either transcribed, binds to regulatory proteins, or is associated with some other biochemical activity.

It however remains controversial whether all of this biochemical activity contributes to cell physiology, or whether a substantial portion of this is the result transcriptional and biochemical noise, which must be actively filtered out by the organism. Excluding protein-coding sequences, introns, and regulatory regions, much of the non-coding DNA is composed of: Many DNA sequences that do not play a role in gene expression have important biological functions. Comparative genomics studies indicate that about 5% of the genome contains sequences of noncoding DNA that are highly conserved, sometimes on time-scales representing hundreds of millions of years, implying that these noncoding regions are under strong evolutionary pressure and positive selection.

Many of these sequences regulate the structure of chromosomes by limiting the regions of heterochromatin formation and regulating structural features of the chromosomes, such as the telomeres and centromeres. Other noncoding regions serve as origins of DNA replication. Finally several regions are transcribed into functional noncoding RNA that regulate the expression of protein-coding genes (for example), mRNA translation and stability, chromatin structure (including histone modifications, for example), DNA methylation (for example), DNA recombination (for example), and cross-regulate other noncoding RNAs (for example). It is also likely that many transcribed noncoding regions do not serve any role and that this transcription is the product of non-specific RNA Polymerase activity.

Pseudogenes

Pseudogenes are inactive copies of protein-coding genes, often generated by gene duplication, that have become nonfunctional through the accumulation of inactivating mutations. Table 1 shows that the number of pseudogenes in the human genome is on the order of 13,000, and in some chromosomes is nearly the same as the number of functional protein-coding genes. Gene duplication is a major mechanism through which new genetic material is generated during molecular evolution.

For example, the olfactory receptor gene family is one of the best-documented examples of pseudogenes in the human genome. More than 60 percent of the genes in this family are non-functional pseudogenes in humans. By comparison, only 20 percent of genes in the mouse olfactory receptor gene family are pseudogenes. Research suggests that this is a species-specific characteristic, as the most closely related primates all have proportionally fewer pseudogenes. This genetic discovery helps to explain the less acute sense of smell in humans relative to other mammals.

Genes for Noncoding RNA (ncRNA)

Noncoding RNA molecules play many essential roles in cells, especially in the many reactions of protein synthesis and RNA processing. Noncoding RNA include tRNA, ribosomal RNA, microRNA, snRNA and other non-coding RNA genes including about 60,000 long non coding RNAs (lncRNAs). It should be noted that while the number of reported lncRNA genes continues to rise and the exact number in the human genome is yet to be defined, many of them are argued to be non-functional.

Many ncRNAs are critical elements in gene regulation and expression. Noncoding RNA also contributes to epigenetics, transcription, RNA splicing, and the translational machinery. The role of RNA in genetic regulation and disease offers a new potential level of unexplored genomic complexity.

Introns and Untranslated Regions of mRNA

In addition to the ncRNA molecules that are encoded by discrete genes, the initial transcripts of protein coding genes usually contain extensive noncoding sequences, in the form of introns, 5'-untranslated regions (5'-UTR), and 3'-untranslated regions (3'-UTR). Within most protein-coding genes of the human genome, the length of intron sequences is 10- to 100-times the length of exon sequences (Table 2).

Regulatory DNA Sequences

The human genome has many different regulatory sequences which are crucial to controlling gene expression. Conservative estimates indicate that these sequences make up 8% of the genome, however extrapolations from the ENCODE project give that 20-40% of the genome is gene regulatory sequence. Some types of non-coding DNA are genetic "switches" that do not encode proteins, but do regulate when and where genes are expressed (called enhancers).

Regulatory sequences have been known since the late 1960s. The first identification of regulatory sequences in the human genome relied on recombinant DNA technology. Later with the advent of genomic sequencing, the identification of these sequences could be inferred by evolutionary conservation. The evolutionary branch between the primates and mouse, for example, occurred 70–90 million years ago. So computer comparisons of gene sequences that identify conserved non-coding sequences will be an indication of their importance in duties such as gene regulation.

Other genomes have been sequenced with the same intention of aiding conservation-guided methods, for exampled the pufferfish genome. However, regulatory sequences disappear and re-evolve during evolution at a high rate.

As of 2012, the efforts have shifted toward finding interactions between DNA and regulatory proteins by the technique ChIP-Seq, or gaps where the DNA is not packaged by histones (DNase hypersensitive sites), both of which tell where there are active regulatory sequences in the investigated cell type.

Repetitive DNA Sequences

Repetitive DNA sequences comprise approximately 50% of the human genome.

About 8% of the human genome consists of tandem DNA arrays or tandem repeats, low complexity repeat sequences that have multiple adjacent copies (e.g. "CAGCAGCAG..."). The tandem sequences may be of variable lengths, from two nucleotides to tens of nucleotides. These sequences are highly variable, even among closely related individuals, and so are used for genealogical DNA testing and forensic DNA analysis.

Repeated sequences of fewer than ten nucleotides (e.g. the dinucleotide repeat $(AC)_n$) are termed microsatellite sequences. Among the microsatellite sequences, trinucleotide repeats are of particular importance, as sometimes occur within coding regions of genes for proteins and may lead to genetic disorders. For example, Huntington's disease results from an expansion of the trinucleotide repeat $(CAG)_n$ within the *Huntingtin* gene on human chromosome 4. Telomeres (the ends of linear chromosomes) end with a microsatellite hexanucleotide repeat of the sequence $(TTAGGG)_n$.

Tandem repeats of longer sequences (arrays of repeated sequences 10–60 nucleotides long) are termed minisatellites.

Mobile Genetic Elements (Transposons) and their Relics

Transposable genetic elements, DNA sequences that can replicate and insert copies of themselves at other locations within a host genome, are an abundant component in the human genome. The most abundant transposon lineage, *Alu*, has about 50,000 active copies, and can be inserted into intragenic and intergenic regions. One other lineage, LINE-1, has about 100 active copies per genome (the number varies between people). Together with non-functional relics of old transposons, they account for over half of total human DNA. Sometimes called "jumping genes", transposons have played a major role in sculpting the human genome. Some of these sequences represent endogenous retroviruses, DNA copies of viral sequences that have become permanently integrated into the genome and are now passed on to succeeding generations.

Mobile elements within the human genome can be classified into LTR retrotransposons (8.3% of total genome), SINEs (13.1% of total genome) including Alu elements, LINEs (20.4% of total genome), SVAs and Class II DNA transposons (2.9% of total genome).

Genomic Variation in Humans

Human Reference Genome

With the exception of identical twins, all humans show significant variation in genomic DNA sequences. The Human Reference Genome (HRG) is used as a standard sequence reference.

There are several important points concerning the Human Reference Genome:

- The HRG is a haploid sequence. Each chromosome is represented once.
- The HRG is a composite sequence, and does not correspond to any actual human individual.
- The HRG is periodically updated to correct errors and ambiguities.
- The HRG in no way represents an "ideal" or "perfect" human individual. It is simply a standardized representation or model that is used for comparative purposes.

Measuring Human Genetic Variation

Most studies of human genetic variation have focused on single-nucleotide polymorphisms (SNPs), which are substitutions in individual bases along a chromosome. Most analyses estimate that SNPs occur 1 in 1000 base pairs, on average, in the euchromatic human genome, although they do not occur at a uniform density. Thus follows the popular statement that "we are all, regardless of race, genetically 99.9% the same", although this would be somewhat qualified by most geneticists. For example, a much larger fraction of the genome is now thought to be involved in copy number variation. A large-scale collaborative effort to catalog SNP variations in the human genome is being undertaken by the International HapMap Project.

The genomic loci and length of certain types of small repetitive sequences are highly variable from person to person, which is the basis of DNA fingerprinting and DNA paternity testing technologies. The heterochromatic portions of the human genome, which total several hundred million base pairs, are also thought to be quite variable within the human population (they are so repetitive and so long that they cannot be accurately sequenced with current technology). These regions contain few genes, and it is unclear whether any significant phenotypic effect results from typical variation in repeats or heterochromatin.

Most gross genomic mutations in gamete germ cells probably result in inviable embryos; however, a number of human diseases are related to large-scale genomic abnormalities. Down syndrome, Turner Syndrome, and a number of other diseases result from nondisjunction of entire chromosomes. Cancer cells frequently have aneuploidy of chromosomes and chromosome arms, although a cause and effect relationship between aneuploidy and cancer has not been established.

Mapping Human Genomic Variation

Whereas a genome sequence lists the order of every DNA base in a genome, a genome map identifies the landmarks. A genome map is less detailed than a genome sequence and aids in navigating around the genome.

An example of a variation map is the HapMap being developed by the International HapMap Project. The HapMap is a haplotype map of the human genome, "which will describe the common patterns of human DNA sequence variation." It catalogs the patterns of small-scale variations in the genome that involve single DNA letters, or bases.

Researchers published the first sequence-based map of large-scale structural variation across the human genome in the journal *Nature* in May 2008. Large-scale structural variations are differences in the genome among people that range from a few thousand to a few million DNA bases; some are gains or losses of stretches of genome sequence and others appear as re-arrangements of stretches of sequence. These variations include differences in the number of copies individuals have of a particular gene, deletions, translocations and inversions.

SNP Frequency Across the Human Genome

Single-nucleotide polymorphisms (SNPs) do not occur homogeneously across the human genome. In fact, there is enormous diversity in SNP frequency between genes, reflecting different selective pressures on each gene as well as different mutation and recombination rates across the genome.

However, studies on SNPs are biased towards coding regions, the data generated from them are unlikely to reflect the overall distribution of SNPs throughout the genome. Therefore, the SNP Consortium protocol was designed to identify SNPs with no bias towards coding regions and the Consortium's 100,000 SNPs generally reflect sequence diversity across the human chromosomes. The SNP Consortium aims to expand the number of SNPs identified across the genome to 300 000 by the end of the first quarter of 2001.

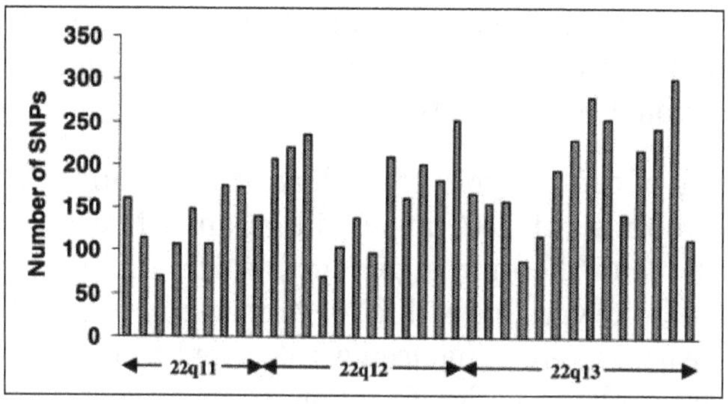

TSC SNP distribution along the long arm of chromosome 22. Each column represents a 1 Mb interval; the approximate cytogenetic position is given on the x-axis. Clear peaks and troughs of SNP density can be seen, possibly reflecting different rates of mutation, recombination and selection.

Changes in non-coding sequence and synonymous changes in coding sequence are generally more common than non-synonymous changes, reflecting greater selective pressure reducing diversity at positions dictating amino acid identity. Transitional changes are more common than transversions, with CpG dinucleotides showing the highest mutation rate, presumably due to deamination.

Personal Genomes

A personal genome sequence is a (nearly) complete sequence of the chemical base pairs that make up the DNA of a single person. Because medical treatments have different effects on different people due to genetic variations such as single-nucleotide polymorphisms (SNPs), the analysis of personal genomes may lead to personalized medical treatment based on individual genotypes.

The first personal genome sequence to be determined was that of Craig Venter in 2007. Personal genomes had not been sequenced in the public Human Genome Project to protect the identity of volunteers who provided DNA samples. That sequence was derived from the DNA of several volunteers from a diverse population. However, early in the Venter-led Celera Genomics genome sequencing effort the decision was made to switch from sequencing a composite sample to using DNA from a single individual, later revealed to have been Venter himself. Thus the Celera human genome sequence released in 2000 was largely that of one man. Subsequent replacement of the early composite-derived data and determination of the diploid sequence, representing both sets of chromosomes, rather than a haploid sequence originally reported, allowed the release of the first personal genome. In April 2008, that of James Watson was also completed. Since then hundreds of personal genome sequences have been released, including those of Desmond Tutu, and of a Paleo-Eskimo. In November 2013, a Spanish family made their personal genomics data obtained by direct-to-consumer genetic testing with 23andMe publicly available under a Creative Commons public domain license. This is believed to be the first such public genomics dataset for a whole family.

The sequencing of individual genomes further unveiled levels of genetic complexity that had not been appreciated before. Personal genomics helped reveal the significant level of diversity in the human genome attributed not only to SNPs but structural variations as well. However, the application of such knowledge to the treatment of disease and in the medical field is only in its very beginnings. Exome sequencing has become increasingly popular as a tool to aid in diagnosis of genetic disease because the exome contributes only 1% of the genomic sequence but accounts for roughly 85% of mutations that contribute significantly to disease.

Human Genetic Disorders

Most aspects of human biology involve both genetic (inherited) and non-genetic (environmental) factors. Some inherited variation influences aspects of our biology that are not medical in nature (height, eye color, ability to taste or smell certain compounds, etc.). Moreover, some genetic disorders only cause disease in combination with the appropriate environmental factors (such as diet). With these caveats, genetic disorders may be described as clinically defined diseases caused by genomic DNA sequence variation. In the most straightforward cases, the disorder can be associated with variation in a single gene. For example, cystic fibrosis is caused by mutations in the CFTR gene, and is the most common recessive disorder in caucasian populations with over 1,300 different mutations known.

Disease-causing mutations in specific genes are usually severe in terms of gene function, and are fortunately rare, thus genetic disorders are similarly individually rare. However, since there are many genes that can vary to cause genetic disorders, in aggregate they constitute a significant component of known medical conditions, especially in pediatric medicine. Molecularly characterized genetic disorders are those for which the underlying causal gene has been identified, currently there are approximately 2,200 such disorders annotated in the OMIM database.

Studies of genetic disorders are often performed by means of family-based studies. In some instances population based approaches are employed, particularly in the case of so-called founder populations such as those in Finland, French-Canada, Utah, Sardinia, etc. Diagnosis and treatment of genetic disorders are usually performed by a geneticist-physician trained in clinical/medical genetics. The results of the Human Genome Project are likely to provide increased availability of genetic testing for gene-related disorders, and eventually improved treatment. Parents can be screened for hereditary conditions and counselled on the consequences, the probability it will be inherited, and how to avoid or ameliorate it in their offspring.

As noted above, there are many different kinds of DNA sequence variation, ranging from complete extra or missing chromosomes down to single nucleotide changes. It is generally presumed that much naturally occurring genetic variation in human populations is phenotypically neutral, i.e. has little or no detectable effect on the physiology of the individual (although there may be fractional differences in fitness defined over evolutionary time frames). Genetic disorders can be caused by any or all known types of sequence variation. To molecularly characterize a new genetic disorder, it is necessary to establish a causal link between a particular genomic sequence variant and the clinical disease under investigation. Such studies constitute the realm of human molecular genetics.

With the advent of the Human Genome and International HapMap Project, it has become feasible to explore subtle genetic influences on many common disease conditions such as diabetes, asthma, migraine, schizophrenia, etc. Although some causal links have been made between genomic sequence variants in particular genes and some of these diseases, often with much publicity in the general media, these are usually not considered to be genetic disorders *per se* as their causes are complex, involving many different genetic and environmental factors. Thus there may be disagreement in particular cases whether a specific medical condition should be termed a genetic disorder. The categorized table below provides the prevalence as well as the genes or chromosomes associated with some human genetic disorders.

Disorder	Prevalence	Chromosome or gene involved
	Chromosomal conditions	
Down syndrome	1:600	Chromosome 21
Klinefelter syndrome	1:500–1000 males	Additional X chromosome
Turner syndrome	1:2000 females	Loss of X chromosome
Sickle cell anemia	1 in 50 births in parts of Africa; rarer elsewhere	β-globin (on chromosome 11)
	Cancers	
Breast/Ovarian cancer (susceptibility)	~5% of cases of these cancer types	BRCA1, BRCA2
FAP (hereditary nonpolyposis coli)	1:3500	APC
Lynch syndrome	5–10% of all cases of bowel cancer	MLH1, MSH2, MSH6, PMS2
	Neurological conditions	
Huntington disease	1:20000	Huntingtin
Alzheimer disease - early onset	1:2500	PS1, PS2, APP
	Other conditions	
Cystic fibrosis	1:2500	CFTR
Duchenne muscular dystrophy	1:3500 boys	Dystrophin

Evolution

Comparative genomics studies of mammalian genomes suggest that approximately 5% of the human genome has been conserved by evolution since the divergence of extant lineages approximately 200 million years ago, containing the vast majority of genes. The published chimpanzee genome differs from that of the human genome by 1.23% in direct sequence comparisons. Around 20% of this figure is accounted for by variation within each species, leaving only ~1.06% consistent sequence divergence between humans and chimps at shared genes. This nucleotide by nucleotide difference is dwarfed, however, by the portion of each genome that is not shared, including around 6% of functional genes that are unique to either humans or chimps.

In other words, the considerable observable differences between humans and chimps may be due as much or more to genome level variation in the number, function and expression of genes rather than DNA sequence changes in shared genes. Indeed, even within humans, there has been found to be a previously unappreciated amount of copy number variation (CNV) which can make up as much as 5 – 15% of the human genome. In other words, between humans, there could be +/-

500,000,000 base pairs of DNA, some being active genes, others inactivated, or active at different levels. The full significance of this finding remains to be seen. On average, a typical human protein-coding gene differs from its chimpanzee ortholog by only two amino acid substitutions; nearly one third of human genes have exactly the same protein translation as their chimpanzee orthologs. A major difference between the two genomes is human chromosome 2, which is equivalent to a fusion product of chimpanzee chromosomes 12 and 13. (later renamed to chromosomes 2A and 2B, respectively).

Humans have undergone an extraordinary loss of olfactory receptor genes during our recent evolution, which explains our relatively crude sense of smell compared to most other mammals. Evolutionary evidence suggests that the emergence of color vision in humans and several other primate species has diminished the need for the sense of smell.

Mitochondrial DNA

The human mitochondrial DNA is of tremendous interest to geneticists, since it undoubtedly plays a role in mitochondrial disease. It also sheds light on human evolution; for example, analysis of variation in the human mitochondrial genome has led to the postulation of a recent common ancestor for all humans on the maternal line of descent.

Due to the lack of a system for checking for copying errors, mitochondrial DNA (mtDNA) has a more rapid rate of variation than nuclear DNA. This 20-fold higher mutation rate allows mtDNA to be used for more accurate tracing of maternal ancestry. Studies of mtDNA in populations have allowed ancient migration paths to be traced, such as the migration of Native Americans from Siberia or Polynesians from southeastern Asia. It has also been used to show that there is no trace of Neanderthal DNA in the European gene mixture inherited through purely maternal lineage. Due to the restrictive all or none manner of mtDNA inheritance, this result (no trace of Neanderthal mtDNA) would be likely unless there were a large percentage of Neanderthal ancestry, or there was strong positive selection for that mtDNA (for example, going back 5 generations, only 1 of your 32 ancestors contributed to your mtDNA, so if one of these 32 was pure Neanderthal you would expect that ~3% of your autosomal DNA would be of Neanderthal origin, yet you would have a ~97% chance to have no trace of Neanderthal mtDNA).

Epigenome

Epigenetics describes a variety of features of the human genome that transcend its primary DNA sequence, such as chromatin packaging, histone modifications and DNA methylation, and which are important in regulating gene expression, genome replication and other cellular processes. Epigenetic markers strengthen and weaken transcription of certain genes but do not affect the actual sequence of DNA nucleotides. DNA methylation is a major form of epigenetic control over gene expression and one of the most highly studied topics in epigenetics. During development, the human DNA methylation profile experiences dramatic changes. In early germ line cells, the genome has very low methylation levels. These low levels generally describe active genes. As development progresses, parental imprinting tags lead to increased methylation activity.

Epigenetic patterns can be identified between tissues within an individual as well as between individuals themselves. Identical genes that have differences only in their epigenetic state are called epialleles. Epialleles can be placed into three categories: those directly determined by an individual's genotype, those influenced by genotype, and those entirely independent of genotype. The epigenome is also influenced significantly by environmental factors. Diet, toxins, and hormones impact the epigenetic state. Studies in dietary manipulation have demonstrated that methyl-deficient diets are associated with hypomethylation of the epigenome. Such studies establish epigenetics as an important interface between the environment and the genome.

Genetic Disorder

A genetic disorder is a genetic problem caused by one or more abnormalities in the genome, especially a condition that is present from birth (congenital). Most genetic disorders are quite rare and affect one person in every several thousands or millions.

Genetic disorders may be hereditary, passed down from the parents' genes. In other genetic disorders, defects may be caused by new mutations or changes to the DNA. In such cases, the defect will only be passed down if it occurs in the germ line. The same disease, such as some forms of cancer, may be caused by an inherited genetic condition in some people, by new mutations in other people, and mainly by environmental causes in other people. Whether, when and to what extent a person with the genetic defect or abnormality will actually suffer from the disease is almost always affected by the environmental factors and events in the person's development.

Some types of recessive gene disorders confer an advantage in certain environments when only one copy of the gene is present.

Single-gene

Prevalence of some single-gene disorders	
Disorder prevalence (approximate)	
Autosomal dominant	
Familial hypercholesterolemia	1 in 500
Polycystic kidney disease	1 in 1250
Neurofibromatosis type I	1 in 2,500
Hereditary spherocytosis	1 in 5,000
Marfan syndrome	1 in 4,000
Huntington's disease	1 in 15,000
Autosomal recessive	
Sickle cell anaemia	1 in 625
Cystic fibrosis	1 in 2,000
Tay-Sachs disease	1 in 3,000
Phenylketonuria	1 in 12,000
Mucopolysaccharidoses	1 in 25,000
Lysosomal acid lipase deficiency	1 in 40,000
Glycogen storage diseases	1 in 50,000

Galactosemia	1 in 57,000
X-linked	
Duchenne muscular dystrophy	1 in 7,000
Hemophilia	1 in 10,000
Values are for liveborn infants	

A single-gene disorder is the result of a single mutated gene. Over 4000 human diseases are caused by single-gene defects. Single-gene disorders can be passed on to subsequent generations in several ways. Genomic imprinting and uniparental disomy, however, may affect inheritance patterns. The divisions between recessive and dominant types are not "hard and fast", although the divisions between autosomal and X-linked types are (since the latter types are distinguished purely based on the chromosomal location of the gene). For example, achondroplasia is typically considered a dominant disorder, but children with two genes for achondroplasia have a severe skeletal disorder of which achondroplasics could be viewed as carriers. Sickle-cell anemia is also considered a recessive condition, but heterozygous carriers have increased resistance to malaria in early childhood, which could be described as a related dominant condition. When a couple where one partner or both are sufferers or carriers of a single-gene disorder wish to have a child, they can do so through *in vitro* fertilization, which means they can then have a preimplantation genetic diagnosis to check whether the embryo has the genetic disorder.

Autosomal Dominant

Only one mutated copy of the gene will be necessary for a person to be affected by an autosomal dominant disorder. Each affected person usually has one affected parent. The chance a child will inherit the mutated gene is 50%. Autosomal dominant conditions sometimes have reduced penetrance, which means although only one mutated copy is needed, not all individuals who inherit that mutation go on to develop the disease. Examples of this type of disorder are Huntington's disease, neurofibromatosis type 1, neurofibromatosis type 2, Marfan syndrome, hereditary nonpolyposis colorectal cancer, hereditary multiple exostoses (a highly penetrant autosomal dominant disorder), Tuberous sclerosis, Von Willebrand disease, and acute intermittent porphyria. Birth defects are also called congenital anomalies.

Autosomal Recessive

Two copies of the gene must be mutated for a person to be affected by an autosomal recessive disorder. An affected person usually has unaffected parents who each carry a single copy of the mutated gene (and are referred to as carriers). Two unaffected people who each carry one copy of the mutated gene have a 25% risk with each pregnancy of having a child affected by the disorder. Examples of this type of disorder are Albinism, Medium-chain acyl-CoA dehydrogenase deficiency, cystic fibrosis, sickle-cell disease, Tay-Sachs disease, Niemann-Pick disease, spinal muscular atrophy, and Roberts syndrome. Certain other phenotypes, such as wet versus dry earwax, are also determined in an autosomal recessive fashion.

X-linked Dominant

X-linked dominant disorders are caused by mutations in genes on the X chromosome. Only a few disorders have this inheritance pattern, with a prime example being X-linked hypophosphatemic rickets. Males and females are both affected in these disorders, with males typically being more severely affected than females. Some X-linked dominant conditions, such as Rett syndrome, incontinentia pigmenti type 2, and Aicardi syndrome, are usually fatal in males either *in utero* or shortly after birth, and are therefore predominantly seen in females. Exceptions to this finding are extremely rare cases in which boys with Klinefelter syndrome (47,XXY) also inherit an X-linked dominant condition and exhibit symptoms more similar to those of a female in terms of disease severity. The chance of passing on an X-linked dominant disorder differs between men and women. The sons of a man with an X-linked dominant disorder will all be unaffected (since they receive their father's Y chromosome), and his daughters will all inherit the condition. A woman with an X-linked dominant disorder has a 50% chance of having an affected fetus with each pregnancy, although it should be noted that in cases such as incontinentia pigmenti, only female offspring are generally viable. In addition, although these conditions do not alter fertility *per se*, individuals with Rett syndrome or Aicardi syndrome rarely reproduce.

X-linked Recessive

X-linked recessive conditions are also caused by mutations in genes on the X chromosome. Males are more frequently affected than females, and the chance of passing on the disorder differs between men and women. The sons of a man with an X-linked recessive disorder will not be affected, and his daughters will carry one copy of the mutated gene. A woman who is a carrier of an X-linked recessive disorder (X^RX^r) has a 50% chance of having sons who are affected and a 50% chance of having daughters who carry one copy of the mutated gene and are therefore carriers. X-linked recessive conditions include the serious diseases hemophilia A, Duchenne muscular dystrophy, and Lesch-Nyhan syndrome, as well as common and less serious conditions such as male pattern baldness and red-green color blindness. X-linked recessive conditions can sometimes manifest in females due to skewed X-inactivation or monosomy X (Turner syndrome).

Y-linked

Y-linked disorders, also called holandric disorders, are caused by mutations on the Y chromosome. These conditions display may only be transmitted from the heterogametic sex (e.g. male humans) to offspring of the same sex. More simply, this means that Y-linked disorders in humans can only be passed from men to their sons; females can never be affected because they do not possess Y-allosomes.

Y-linked disorders are exceedingly rare but the most well-known examples typically cause infertility. Reproduction in such conditions is only possible through the circumvention of infertility by medical intervention.

Mitochondrial

This type of inheritance, also known as maternal inheritance, applies to genes in mitochondrial DNA. Because only egg cells contribute mitochondria to the developing embryo, only mothers can pass on mitochondrial conditions to their children. An example of this type of disorder is Leber's hereditary optic neuropathy.

Many Genes

Genetic disorders may also be complex, multifactorial, or polygenic, meaning they are likely associated with the effects of multiple genes in combination with lifestyles and environmental factors. Multifactorial disorders include heart disease and diabetes. Although complex disorders often cluster in families, they do not have a clear-cut pattern of inheritance. This makes it difficult to determine a person's risk of inheriting or passing on these disorders. Complex disorders are also difficult to study and treat, because the specific factors that cause most of these disorders have not yet been identified. Studies which aim to identify the cause of complex disorders can use several methodological approaches to determine genotype-phenotype associations. One method, the genotype-first approach, starts by identifying genetic variants within patients and then determining the associated clinical manifestations. This is opposed to the more traditional phenotype-first approach, and may identify causal factors that have previously been obscured by clinical heterogeneity, penetrance, and expressivity.

On a pedigree, polygenic diseases do tend to "run in families", but the inheritance does not fit simple patterns as with Mendelian diseases. But this does not mean that the genes cannot eventually be located and studied. There is also a strong environmental component to many of them (e.g., blood pressure):

- asthma.
- autoimmune diseases such as multiple sclerosis.
- cancers.
- ciliopathies.
- cleft palate.
- diabetes.
- heart disease.
- hypertension.
- inflammatory bowel disease.
- intellectual disability.
- mood disorder.
- obesity.
- refractive error.
- infertility.

Diagnosis

Due to the wide range of genetic disorders that are presently known, diagnosis of a genetic disorder is widely varied and dependent of the disorder. Most genetic disorders are diagnosed at birth or during early childhood, however some, such as Huntington's disease, can escape detection until the patient is well into adulthood.

The basic aspects of a genetic disorder rests on the inheritance of genetic material. With an in depth family history, it is possible to anticipate possible disorders in children which direct medical professionals to specific tests depending on the disorder and allow parents the chance to prepare for potential lifestyle changes, anticipate the possibility of stillbirth, or contemplate termination. Prenatal diagnosis can detect the presence of characteristic abnormalities in fetal development through ultrasound, or detect the presence of characteristic substances via invasive procedures which involve inserting probes or needles into the uterus such as in amniocentesis.

Prognosis

Not all genetic disorders directly result in death, however there are no known cures for genetic disorders. Many genetic disorders affect stages of development such as Down's Syndrome. While others result in purely physical symptoms such as Muscular Dystrophy. Other disorders, such as Huntington's Disease show no signs until adulthood. During the active time of a genetic disorder, patients mostly rely on maintaining or slowing the degradation of quality of life and maintain patient autonomy. This includes physical therapy, pain management, and may include a selection of alternative medicine programs.

Treatment

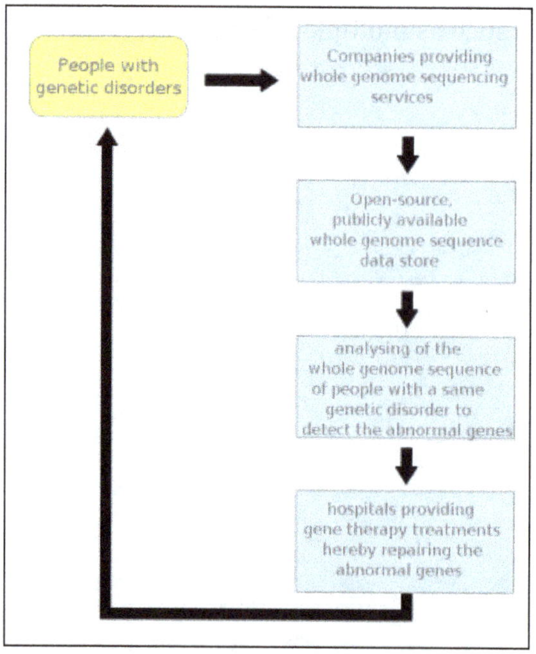

From personal genomics to gene therapy.

The treatment of genetic disorders is an ongoing battle with over 1800 gene therapy clinical trials having been completed, are ongoing, or have been approved worldwide. Despite this, most treatment options revolve around treating the symptoms of the disorders in an attempt to improve patient quality of life.

Gene therapy refers to a form of treatment where a healthy gene is introduced to a patient. This should alleviate the defect caused by a faulty gene or slow the progression of disease. A major obstacle has been the delivery of genes to the appropriate cell, tissue, and organ affected by the disorder. How does one introduce a gene into the potentially trillions of cells which carry the defective copy? This question has been the roadblock between understanding the genetic disorder and correcting the genetic disorder.

Gene Therapy

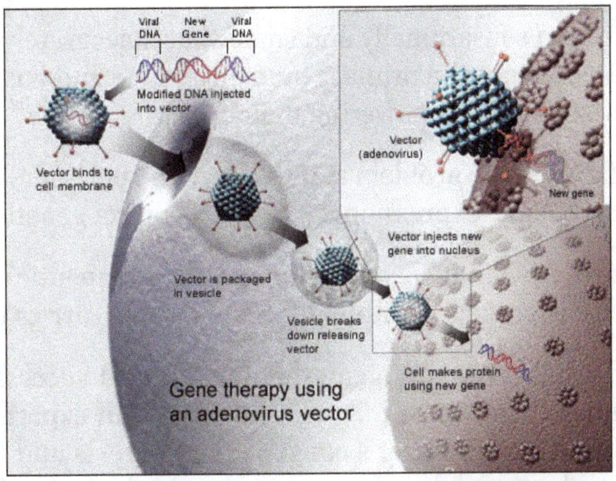

Gene therapy using an adenovirus vector. In some cases, the adenovirus will insert the new gene into a cell. If the treatment is successful, the new gene will make a functional protein to treat a disease.

Gene therapy is the therapeutic delivery of nucleic acid polymers into a patient's cells as a drug to treat disease.

The first attempt at modifying human DNA was performed in 1980 by Martin Cline, but the first successful and approved nuclear gene transfer in humans was performed in May 1989. The first therapeutic use of gene transfer as well as the first direct insertion of human DNA into the nuclear genome was performed by French Anderson in a trial starting in September 1990.

Between 1989 and February 2016, over 2,300 clinical trials had been conducted, more than half of them in phase I.

It should be noted that not all medical procedures that introduce alterations to a patient's genetic makeup can be considered gene therapy. Bone marrow transplantation and organ transplants in general have been found to introduce foreign DNA into patients. Gene therapy is defined by the precision of the procedure and the intention of direct therapeutic effects.

Background

Gene therapy was conceptualized in 1972, by authors who urged caution before commencing human gene therapy studies.

The first attempt, an unsuccessful one, at gene therapy (as well as the first case of medical transfer of foreign genes into humans not counting organ transplantation) was performed by Martin Cline on 10 July 1980. Cline claimed that one of the genes in his patients was active six months later, though he never published this data or had it verified and even if he is correct, it's unlikely it produced any significant beneficial effects treating beta-thalassemia.

After extensive research on animals throughout the 1980s and a 1989 bacterial gene tagging trial on humans, the first gene therapy widely accepted as a success was demonstrated in a trial that started on September 14, 1990, when Ashi DeSilva was treated for ADA-SCID.

The first somatic treatment that produced a permanent genetic change was performed in 1993.

This procedure was referred to sensationally and somewhat inaccurately in the media as a "three parent baby", though mtDNA is not the primary human genome and has little effect on an organism's individual characteristics beyond powering their cells.

Gene therapy is a way to fix a genetic problem at its source. The polymers are either translated into proteins, interfere with target gene expression, or possibly correct genetic mutations.

The most common form uses DNA that encodes a functional, therapeutic gene to replace a mutated gene. The polymer molecule is packaged within a "vector", which carries the molecule inside cells.

Early clinical failures led to dismissals of gene therapy. Clinical successes since 2006 regained researchers' attention, although as of 2014, it was still largely an experimental technique. These include treatment of retinal diseases Leber's congenital amaurosis and choroideremia, X-linked SCID, ADA-SCID, adrenoleukodystrophy, chronic lymphocytic leukemia (CLL), acute lymphocytic leukemia (ALL), multiple myeloma, haemophilia and Parkinson's disease. Between 2013 and April 2014, US companies invested over $600 million in the field.

The first commercial gene therapy, Gendicine, was approved in China in 2003 for the treatment of certain cancers. In 2011 Neovasculgen was registered in Russia as the first-in-class gene-therapy drug for treatment of peripheral artery disease, including critical limb ischemia. In 2012 Glybera, a treatment for a rare inherited disorder, became the first treatment to be approved for clinical use in either Europe or the United States after its endorsement by the European Commission.

Approaches

Following early advances in genetic engineering of bacteria, cells, and small animals, scientists started considering how to apply it to medicine. Two main approaches were considered – replacing or disrupting defective genes. Scientists focused on diseases caused by single-gene defects, such as cystic fibrosis, haemophilia, muscular dystrophy, thalassemia and sickle cell anemia. Glybera treats one such disease, caused by a defect in lipoprotein lipase.

DNA must be administered, reach the damaged cells, enter the cell and express/disrupt a protein. Multiple delivery techniques have been explored. The initial approach incorporated DNA into an engineered virus to deliver the DNA into a chromosome. Naked DNA approaches have also been explored, especially in the context of vaccine development.

Generally, efforts focused on administering a gene that causes a needed protein to be expressed. More recently, increased understanding of nuclease function has led to more direct DNA editing, using techniques such as zinc finger nucleases and CRISPR. The vector incorporates genes into chromosomes. The expressed nucleases then knock out and replace genes in the chromosome. As of 2014 these approaches involve removing cells from patients, editing a chromosome and returning the transformed cells to patients.

Future of CRISPR-Cas 9

Gene editing has been a potential therapy for many genetic diseases. Targeted genome editing using nucleases provides a general method for inducing deletions or insertion. An earlier method for targeting relies on protein-DNA interactions; however, the most recent one, using CRISPR – associated protein 9 (Cas9), provides better specificity, simplicity, speed and pricing. The CRISPR System was first identified in single cell archaea (Prokaryotes). It is now widely used in different cell types and organisms including human cells (HEK293T, HeLa, iPSC), mouse, fruit fly, rice, wheat etc. CRISPR –Cas9 genome editing has potential applications in human gene therapy, screening drug targets, synthetic biology, agriculture, programmable RNA targeting, and viral gene disruption.

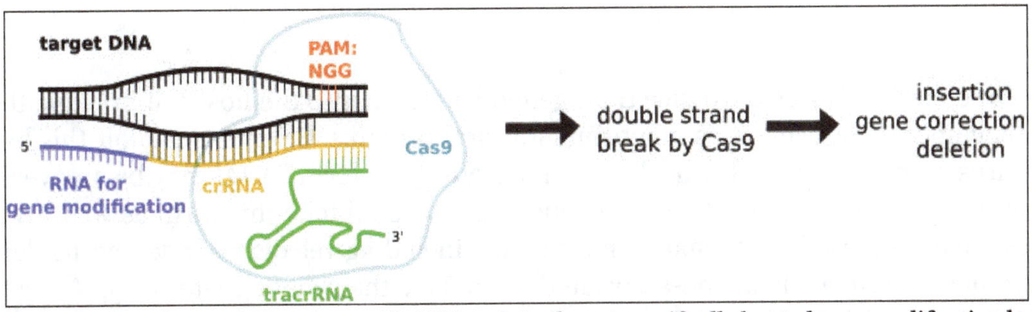

A duplex of crRNA and tracrRNA acts as guide RNA to introduce a specifically located gene modification based on the RNA 5' upstream of the crRNA. Cas9 binds the tracrRNA and needs a DNA binding sequence (5'NGG3'), which is called protospacer adjacent motif (PAM). After binding, Cas9 introduces a DNA double strand break, which is then followed by gene modification via homologous recombination (HDR) or non-homologous end joining (NHEJ).

Other technologies employ antisense, small interfering RNA and other DNA. To the extent that these technologies do not alter DNA, but instead directly interact with molecules such as RNA, they are not considered "gene therapy" per se.

Cell Types

Somatic Cell

In somatic cell gene therapy (SCGT), the therapeutic genes are transferred into any cell other than a gamete, germ cell, gametocyte or undifferentiated stem cell. Any such modifications affect the individual patient only, and are not inherited by offspring. Somatic gene therapy represents mainstream basic and clinical research, in which therapeutic DNA (either integrated in the genome or as an external episome or plasmid) is used to treat disease.

Over 600 clinical trials utilizing SCGT are underway in the US. Most focus on severe genetic disorders, including immunodeficiencies, haemophilia, thalassaemia and cystic fibrosis. Such single gene disorders are good candidates for somatic cell therapy. The complete correction of a genetic disorder or the replacement of multiple genes is not yet possible. Only a few of the trials are in the advanced stages.

Germline

In germline gene therapy (GGT), germ cells (sperm or eggs) are modified by the introduction of functional genes into their genomes. Modifying a germ cell causes all the organism's cells to contain the modified gene. The change is therefore heritable and passed on to later generations. Australia, Canada, Germany, Israel, Switzerland and the Netherlands prohibit GGT for application in human beings, for technical and ethical reasons, including insufficient knowledge about possible risks to future generations and higher risks versus SCGT. The US has no federal controls specifically addressing human genetic modification (beyond FDA regulations for therapies in general).

Vectors

The delivery of DNA into cells can be accomplished by multiple methods. The two major classes are recombinant viruses (sometimes called biological nanoparticles or viral vectors) and naked DNA or DNA complexes (non-viral methods).

Viruses

In order to replicate, viruses introduce their genetic material into the host cell, tricking the host's cellular machinery into using it as blueprints for viral protcins. Scientists exploit this by substituting a virus's genetic material with therapeutic DNA. (The term 'DNA' may be an oversimplification, as some viruses contain RNA, and gene therapy could take this form as well.) A number of viruses have been used for human gene therapy, including retrovirus, adenovirus, lentivirus, herpes simplex, vaccinia and adeno-associated virus. Like the genetic material (DNA or RNA) in viruses, therapeutic DNA can be designed to simply serve as a temporary blueprint that is degraded naturally or (at least theoretically) to enter the host's genome, becoming a permanent part of the host's DNA in infected cells.

Non-viral

Non-viral methods present certain advantages over viral methods, such as large scale production and low host immunogenicity. However, non-viral methods initially produced lower levels of transfection and gene expression, and thus lower therapeutic efficacy. Later technology remedied this deficiency.

Methods for non-viral gene therapy include the injection of naked DNA, electroporation, the gene gun, sonoporation, magnetofection, the use of oligonucleotides, lipoplexes, dendrimers, and inorganic nanoparticles.

Hurdles

Some of the unsolved problems include:

- Short-lived nature – Before gene therapy can become a permanent cure for a condition, the therapeutic DNA introduced into target cells must remain functional and the cells containing the therapeutic DNA must be stable. Problems with integrating therapeutic DNA into the genome and the rapidly dividing nature of many cells prevent it from achieving long-term benefits. Patients require multiple treatments.

- Immune response – Any time a foreign object is introduced into human tissues, the immune system is stimulated to attack the invader. Stimulating the immune system in a way that reduces gene therapy effectiveness is possible. The immune system's enhanced response to viruses that it has seen before reduces the effectiveness to repeated treatments.
- Problems with viral vectors – Viral vectors carry the risks of toxicity, inflammatory responses, and gene control and targeting issues.
- Multigene disorders – Some commonly occurring disorders, such as heart disease, high blood pressure, Alzheimer's disease, arthritis, and diabetes, are affected by variations in multiple genes, which complicate gene therapy.
- Some therapies may breach the Weismann barrier (between soma and germ-line) protecting the testes, potentially modifying the germline, falling afoul of regulations in countries that prohibit the latter practice.
- Insertional mutagenesis – If the DNA is integrated in a sensitive spot in the genome, for example in a tumor suppressor gene, the therapy could induce a tumor. This has occurred in clinical trials for X-linked severe combined immunodeficiency (X-SCID) patients, in which hematopoietic stem cells were transduced with a corrective transgene using a retrovirus, and this led to the development of T cell leukemia in 3 of 20 patients. One possible solution is to add a functional tumor suppressor gene to the DNA to be integrated. This may be problematic since the longer the DNA is, the harder it is to integrate into cell genomes. CRISPR technology allows researchers to make much more precise genome changes at exact locations.
- Cost – Alipogene tiparvovec or Glybera, for example, at a cost of $1.6 million per patient, was reported in 2013 to be the world's most expensive drug.

Deaths

Three patients' deaths have been reported in gene therapy trials, putting the field under close scrutiny. The first was that of Jesse Gelsinger in 1999. One X-SCID patient died of leukemia in 2003. In 2007, a rheumatoid arthritis patient died from an infection; the subsequent investigation concluded that the death was not related to gene therapy.

History

1970s and Earlier

In 1972 Friedmann and Roblin authored a paper in *Science* titled "Gene therapy for human genetic disease?" Rogers (1970) was cited for proposing that *exogenous good DNA* be used to replace the defective DNA in those who suffer from genetic defects.

1980s

In 1984 a retrovirus vector system was designed that could efficiently insert foreign genes into mammalian chromosomes.

1990s

The first approved gene therapy clinical research in the US took place on 14 September 1990, at the National Institutes of Health (NIH), under the direction of William French Anderson. Four-year-old Ashanti DeSilva received treatment for a genetic defect that left her with ADA-SCID, a severe immune system deficiency. The effects were temporary, but successful.

Cancer gene therapy was introduced in 1992/93 (Trojan et al. 1993). The treatment of glioblastoma multiforme, the malignant brain tumor whose outcome is always fatal, was done using a vector expressing antisense IGF-I RNA (clinical trial approved by NIH n° 1602, and FDA in 1994). This therapy also represents the beginning of cancer immunogene therapy, a treatment which proves to be effective due to the anti-tumor mechanism of IGF-I antisense, which is related to strong immune and apoptotic phenomena.

In 1992 Claudio Bordignon, working at the Vita-Salute San Raffaele University, performed the first gene therapy procedure using hematopoietic stem cells as vectors to deliver genes intended to correct hereditary diseases. In 2002 this work led to the publication of the first successful gene therapy treatment for adenosine deaminase-deficiency (SCID). The success of a multi-center trial for treating children with SCID (severe combined immune deficiency or "bubble boy" disease) from 2000 and 2002, was questioned when two of the ten children treated at the trial's Paris center developed a leukemia-like condition. Clinical trials were halted temporarily in 2002, but resumed after regulatory review of the protocol in the US, the United Kingdom, France, Italy and Germany.

In 1993 Andrew Gobea was born with SCID following prenatal genetic screening. Blood was removed from his mother's placenta and umbilical cord immediately after birth, to acquire stem cells. The allele that codes for adenosine deaminase (ADA) was obtained and inserted into a retrovirus. Retroviruses and stem cells were mixed, after which the viruses inserted the gene into the stem cell chromosomes. Stem cells containing the working ADA gene were injected into Andrew's blood. Injections of the ADA enzyme were also given weekly. For four years T cells (white blood cells), produced by stem cells, made ADA enzymes using the ADA gene. After four years more treatment was needed.

Jesse Gelsinger's death in 1999 impeded gene therapy research in the US. As a result, the FDA suspended several clinical trials pending the reevaluation of ethical and procedural practices.

2000s

The modified cancer gene therapy strategy of antisense IGF-I RNA (NIH n° 1602) using antisense / triple helix anti IGF-I approach was registered in 2002 by Wiley gene therapy clinical trial - n° 635 and 636. The approach has shown promising results in the treatment of six different malignant tumors: glioblastoma, cancers of liver, colon, prostate, uterus and ovary (Collaborative NATO Science Programme on Gene Therapy USA, France, Poland n° LST 980517 conducted by J. Trojan) (Trojan et al., 2012). This anti–gene antisense/triple helix therapy has proven to be efficient, due to the mechanism stopping simultaneously IGF-I expression on translation and transcription levels, strengthening anti-tumor immune and apoptotic phenomena.

2002

Sickle-cell disease can be treated in mice. The mice – which have essentially the same defect that causes human cases – used a viral vector to induce production of fetal hemoglobin (HbF), which normally ceases to be produced shortly after birth. In humans, the use of hydroxyurea to stimulate the production of HbF temporarily alleviates sickle cell symptoms. The researchers demonstrated this treatment to be a more permanent means to increase therapeutic HbF production.

A new gene therapy approach repaired errors in messenger RNA derived from defective genes. This technique has the potential to treat thalassaemia, cystic fibrosis and some cancers.

Researchers created liposomes 25 nanometers across that can carry therapeutic DNA through pores in the nuclear membrane.

2003

In 2003 a research team inserted genes into the brain for the first time. They used liposomes coated in a polymer called polyethylene glycol, which, unlike viral vectors, are small enough to cross the blood–brain barrier.

Short pieces of double-stranded RNA (short, interfering RNAs or siRNAs) are used by cells to degrade RNA of a particular sequence. If a siRNA is designed to match the RNA copied from a faulty gene, then the abnormal protein product of that gene will not be produced.

Gendicine is a cancer gene therapy that delivers the tumor suppressor gene p53 using an engineered adenovirus. In 2003, it was approved in China for the treatment of head and neck squamous cell carcinoma.

2006

In March researchers announced the successful use of gene therapy to treat two adult patients for X-linked chronic granulomatous disease, a disease which affects myeloid cells and damages the immune system. The study is the first to show that gene therapy can treat the myeloid system.

In May a team reported a way to prevent the immune system from rejecting a newly delivered gene. Similar to organ transplantation, gene therapy has been plagued by this problem. The immune system normally recognizes the new gene as foreign and rejects the cells carrying it. The research utilized a newly uncovered network of genes regulated by molecules known as microRNAs. This natural function selectively obscured their therapeutic gene in immune system cells and protected it from discovery. Mice infected with the gene containing an immune-cell microRNA target sequence did not reject the gene.

In August scientists successfully treated metastatic melanoma in two patients using killer T cells genetically retargeted to attack the cancer cells.

In November researchers reported on the use of VRX496, a gene-based immunotherapy for the treatment of HIV that uses a lentiviral vector to deliver an antisense gene against the HIV envelope. In a phase I clinical trial, five subjects with chronic HIV infection who had failed to respond to at least two antiretroviral regimens were treated. A single intravenous infusion of autologous CD4 T cells genetically modified with VRX496 was well tolerated. All patients had stable or decreased viral load; four of the five patients had stable or increased CD4 T cell counts. All five patients had stable or increased immune response to HIV antigens and other pathogens. This was the first evaluation of a lentiviral vector administered in a US human clinical trial.

2007

In May researchers announced the first gene therapy trial for inherited retinal disease. The first operation was carried out on a 23-year-old British male, Robert Johnson, in early 2007.

2008

Leber's congenital amaurosis is an inherited blinding disease caused by mutations in the RPE65 gene. The results of a small clinical trial in children were published in April. Delivery of recombinant adeno-associated virus (AAV) carrying RPE65 yielded positive results. In May two more groups reported positive results in independent clinical trials using gene therapy to treat the condition. In all three clinical trials, patients recovered functional vision without apparent side-effects.

2009

In September researchers were able to give trichromatic vision to squirrel monkeys. In November 2009, researchers halted a fatal genetic disorder called adrenoleukodystrophy in two children using a lentivirus vector to deliver a functioning version of ABCD1, the gene that is mutated in the disorder.

2010s

2010

An April paper reported that gene therapy addressed achromatopsia (color blindness) in dogs by targeting cone photoreceptors. Cone function and day vision were restored for at least 33 months in two young specimens. The therapy was less efficient for older dogs.

In September it was announced that an 18-year-old male patient in France with beta-thalassemia major had been successfully treated. Beta-thalassemia major is an inherited blood disease in which beta haemoglobin is missing and patients are dependent on regular lifelong blood transfusions. The technique used a lentiviral vector to transduce the human ß-globin gene into purified blood and marrow cells obtained from the patient in June 2007. The patient's haemoglobin levels were stable at 9 to 10 g/dL. About a third of the hemoglobin contained the form introduced by the viral vector and blood transfusions were not needed. Further clinical trials were planned. Bone marrow transplants are the only cure for thalassemia, but 75% of patients do not find a matching donor.

2011

In 2007 and 2008, a man was cured of HIV by repeated Hematopoietic stem cell transplantation with double-delta-32 mutation which disables the CCR5 receptor. This cure was accepted by the medical community in 2011. It required complete ablation of existing bone marrow, which is very debilitating.

In August two of three subjects of a pilot study were confirmed to have been cured from chronic lymphocytic leukemia (CLL). The therapy used genetically modified T cells to attack cells that expressed the CD19 protein to fight the disease. In 2013, the researchers announced that 26 of 59 patients had achieved complete remission and the original patient had remained tumor-free.

Human HGF plasmid DNA therapy of cardiomyocytes is being examined as a potential treatment for coronary artery disease as well as treatment for the damage that occurs to the heart after myocardial infarction.

In 2011 Neovasculgen was registered in Russia as the first-in-class gene-therapy drug for treatment of peripheral artery disease, including critical limb ischemia; it delivers the gene encoding for VEGF. Neovasculogen is a plasmid encoding the CMV promoter and the 165 amino acid form of VEGF.

2012

The FDA approved Phase 1 clinical trials on thalassemia major patients in the US for 10 participants in July. The study was expected to continue until 2015.

In July 2012, the European Medicines Agency recommended approval of a gene therapy treatment for the first time in either Europe or the United States. The treatment used Alipogene tiparvovec (Glybera) to compensate for lipoprotein lipase deficiency, which can cause severe pancreatitis. The recommendation was endorsed by the European Commission in November 2012 and commercial rollout began in late 2014.

In December 2012, it was reported that 10 of 13 patients with multiple myeloma were in remission "or very close to it" three months after being injected with a treatment involving genetically engineered T cells to target proteins NY-ESO-1 and LAGE-1, which exist only on cancerous myeloma cells.

2013

In March researchers reported that three of five subjects who had acute lymphocytic leukemia (ALL) had been in remission for five months to two years after being treated with genetically modified T cells which attacked cells with CD19 genes on their surface, i.e. all B-cells, cancerous or not. The researchers believed that the patients' immune systems would make normal T-cells and B-cells after a couple of months. They were also given bone marrow. One patient relapsed and died and one died of a blood clot unrelated to the disease.

Following encouraging Phase 1 trials, in April, researchers announced they were starting Phase 2 clinical trials (called CUPID2 and SERCA-LVAD) on 250 patients at several hospitals to combat heart disease. The therapy was designed to increase the levels of SERCA2, a protein in heart muscles, improving muscle function. The FDA granted this a Breakthrough Therapy Designation to accelerate the trial and approval process. In 2016 it was reported that no improvement was found from the CUPID 2 trial.

In July researchers reported promising results for six children with two severe hereditary diseases had been treated with a partially deactivated lentivirus to replace a faulty gene and after 7–32 months. Three of the children had metachromatic leukodystrophy, which causes children to lose cognitive and motor skills. The other children had Wiskott-Aldrich syndrome, which leaves them to open to infection, autoimmune diseases and cancer. Follow up trials with gene therapy on another six children with Wiskott-Aldrich syndrome were also reported as promising.

In October researchers reported that two children born with adenosine deaminase severe combined immunodeficiency disease (ADA-SCID) had been treated with genetically engineered stem cells 18 months previously and that their immune systems were showing signs of full recovery. Another three children were making progress. In 2014 a further 18 children with ADA-SCID were cured by gene therapy. ADA-SCID children have no functioning immune system and are sometimes known as "bubble children."

Also in October researchers reported that they had treated six haemophilia sufferers in early 2011 using an adeno-associated virus. Over two years later all six were producing clotting factor.

Data from three trials on Topical cystic fibrosis transmembrane conductance regulator gene therapy were reported to not support its clinical use as a mist inhaled into the lungs to treat cystic fibrosis patients with lung infections.

2014

In January researchers reported that six choroideremia patients had been treated with adeno-associated virus with a copy of REP1. Over a six-month to two-year period all had improved their sight. By 2016, 32 patients had been treated with positive results and researchers were hopeful the treatment would be long-lasting. Choroideremia is an inherited genetic eye disease with no approved treatment, leading to loss of sight.

In March researchers reported that 12 HIV patients had been treated since 2009 in a trial with a genetically engineered virus with a rare mutation (CCR5 deficiency) known to protect against HIV with promising results.

Clinical trials of gene therapy for sickle cell disease were started in 2014 although one review failed to find any such trials.

2015

In February LentiGlobin BB305, a gene therapy treatment undergoing clinical trials for treatment of beta thalassemia gained FDA "breakthrough" status after several patients were able to forgo the frequent blood transfusions usually required to treat the disease.

In March researchers delivered a recombinant gene encoding a broadly neutralizing antibody into monkeys infected with simian HIV; the monkeys' cells produced the antibody, which cleared them of HIV. The technique is named immunoprophylaxis by gene transfer (IGT). Animal tests for antibodies to ebola, malaria, influenza and hepatitis are underway.

In March scientists, including an inventor of CRISPR, urged a worldwide moratorium on germline gene therapy, writing "scientists should avoid even attempting, in lax jurisdictions, germline genome modification for clinical application in humans" until the full implications "are discussed among scientific and governmental organizations".

Also in 2015 Glybera was approved for the German market.

In October, researchers announced that they had treated a baby girl, Layla Richards, with an experimental treatment using donor T-cells genetically engineered to attack cancer cells. Two months after the treatment she was still free of her cancer (a highly aggressive form of acute lymphoblastic leukaemia [ALL]). Children with highly aggressive ALL normally have a very poor prognosis and Layla's disease had been regarded as terminal before the treatment.

In December, scientists of major world academies called for a moratorium on inheritable human genome edits, including those related to CRISPR-Cas9 technologies but that basic research including embryo gene editing should continue.

2016

In April the Committee for Medicinal Products for Human Use of the European Medicines Agency endorsed a gene therapy treatment called Strimvelis and recommended it be approved. This treats children born with ADA-SCID and who have no functioning immune system - sometimes called the "bubble baby" disease. This would be the second gene therapy treatment to be approved in Europe.

Fertility

Gene Therapy techniques have the potential to provide alternative treatments for those with infertility. Recently, successful experimentation on mice has proven that fertility can be restored by using the gene therapy method, CRISPR. Spermatogenical stem cells from another organism were transplanted into the testes of an infertile male mouse. The stem cells re-established spermatogenesis and fertility.

Gene Doping

Athletes might adopt gene therapy technologies to improve their performance. Gene doping is not known to occur, but multiple gene therapies may have such effects. Kayser et al. argue that gene doping could level the playing field if all athletes receive equal access. Critics claim that any therapeutic intervention for non-therapeutic/enhancement purposes compromises the ethical foundations of medicine and sports.

Human Genetic Engineering

Genetic engineering could be used to change physical appearance, metabolism, and even improve physical capabilities and mental faculties such as memory and intelligence. Ethical claims about germline engineering include beliefs that every fetus has a right to remain genetically unmodified, that parents hold the right to genetically modify their offspring, and that every child has the right to be born free of preventable diseases. For adults, genetic engineering could be seen as another enhancement technique to add to diet, exercise, education, cosmetics and plastic surgery. Another theorist claims that moral concerns limit but do not prohibit germline engineering.

Possible regulatory schemes include a complete ban, provision to everyone, or professional self-regulation. The American Medical Association's Council on Ethical and Judicial Affairs stated that "genetic interventions to enhance traits should be considered permissible only in severely restricted situations: (1) clear and meaningful benefits to the fetus or child; (2) no trade-off with other characteristics or traits; and (3) equal access to the genetic technology, irrespective of income or other socioeconomic characteristics."

As early in the history of biotechnology as 1990, there have been scientists opposed to attempts to modify the human germline using these new tools, and such concerns have continued as technology progressed. With the advent of new techniques like CRISPR, in March 2015 a group of scientists urged a worldwide moratorium on clinical use of gene editing technologies to edit the human genome in a way that can be inherited. In April 2015, researchers sparked controversy when they reported results of basic research to edit the DNA of non-viable human embryos using CRISPR.

Regulations

Regulations covering genetic modification are part of general guidelines about human-involved biomedical research.

The Helsinki Declaration (Ethical Principles for Medical Research Involving Human Subjects) was amended by the World Medical Association's General Assembly in 2008. This document provides principles physicians and researchers must consider when involving humans as research subjects. The Statement on Gene Therapy Research initiated by the Human Genome Organization (HUGO) in 2001 provides a legal baseline for all countries. HUGO's document emphasizes human freedom and adherence to human rights, and offers recommendations for somatic gene therapy, including the importance of recognizing public concerns about such research.

United States

No federal legislation lays out protocols or restrictions about human genetic engineering. This subject is governed by overlapping regulations from local and federal agencies, including the Department of Health and Human Services, the FDA and NIH's Recombinant DNA Advisory Committee. Researchers seeking federal funds for an investigational new drug application, (commonly the case for somatic human genetic engineering), must obey international and federal guidelines for the protection of human subjects.

NIH serves as the main gene therapy regulator for federally funded research. Privately funded research is advised to follow these regulations. NIH provides funding for research that develops or enhances genetic engineering techniques and to evaluate the ethics and quality in current research. The NIH maintains a mandatory registry of human genetic engineering research protocols that includes all federally funded projects.

An NIH advisory committee published a set of guidelines on gene manipulation. The guidelines discuss lab safety as well as human test subjects and various experimental types that involve genetic changes. Several sections specifically pertain to human genetic engineering, including Section III-C-1. This section describes required review processes and other aspects when seeking approval to begin clinical research involving genetic transfer into a human patient. The protocol for a gene therapy clinical trial must be approved by the NIH's Recombinant DNA Advisory Committee prior to any clinical trial beginning; this is different from any other kind of clinical trial.

As with other kinds of drugs, the FDA regulates the quality and safety of gene therapy products and supervises how these products are used clinically. Therapeutic alteration of the human genome falls under the same regulatory requirements as any other medical treatment. Research involving human subjects, such as clinical trials, must be reviewed and approved by the FDA and an Institutional Review Board.

References

- Milunsky, edited by Aubrey (2004). Genetic disorders and the fetus : diagnosis, prevention, and treatment (5th ed.). Baltimore: Johns Hopkins University Press. ISBN 0801879280
- Rakesh Tamang, Lalji Singh, Kumarasamy Thangaraj (November 2012). "Complex genetic origin of Indian populations and its implications" (PDF). Journal of Biosciences. 37 (5): 911–919. doi:10.1007/s12038-012-9256-9. Retrieved 29, March 2020
- "Chiesi and uniQure delay Glybera launch to add data". Biotechnology. The Pharma Letter. 4 August 2014. Retrieved 28 August 2020
- Barton NH, Briggs DE, Eisen JA, Goldstein DB, Patel NH (2007). Evolution. Cold Spring Harbor, NY: Cold Spring Harbor Laboratory Press. ISBN 0-87969-684-2
- Pollack, Andrew (2 June 2016). "Scientists Announce HGP-Write, Project to Synthesize the Human Genome". New York Times. Retrieved 03, March 2020
- BURGER, LUDWIG; HIRSCHLER, BEN (November 26, 2014). "First gene therapy drug sets million-euro price record". Reuters. Retrieved 12, April 2020
- Griffiths, Anthony J.F.; Wessler, Susan R.; Carroll, Sean B.; Doebley, John (2012). Introduction to Genetic Analysis (10th ed.). New York: W.H. Freeman and Company. p. 58. ISBN 978-1-4292-2943-2
- "Summary of opinion1 (initial authorisation) Strimvelis" (PDF). European Medicines Agency. 1 April 2016. pp. 1–2. Retrieved 01, March 2020
- Priya Moorjani, Kumarasamy Thangaraj, Nick Patterson, Mark Lipson, Po-Ru Loh, Periyasamy Govindaraj, Bonnie Berger, David Reich, Lalji Singh (5 September 2013). "Genetic Evidence for Recent Population Mixture in India" (PDF). American Journal of Human Genetics. 93 (3): 422–438. doi:10.1016/j.ajhg.2013.07.006. PMC 3769933. PMID 23932107. Retrieved 23, July 2020
- Griffiths, Anthony J.F.; Wessler, Susan R.; Carroll, Sean B.; Doebley, John (2012). "2: Single-Gene Inheritance". Introduction to Genetic Analysis (10th ed.). New York: W.H. Freeman and Company. p. 57. ISBN 978-1-4292-2943-2
- Ghosh, Pallab (28 April 2016). "Gene therapy reverses sight loss and is long-lasting". BBC News, Science & Environment. Retrieved 29, June 2020
- Fernàndez-Ruiz, Irene (4 February 2016). "No improvement in outcomes with gene therapy for heart failure". Nature Reviews Cardiology. 13: 122–123. doi:10.1038/nrcardio.2016.14. Retrieved 08, July 2020
- "Gene Therapy Turns Several Leukemia Patients Cancer Free. Will It Work for Other Cancers, Too?". Singularity Hub. Retrieved 14, May 2020

6

Various Types of Genetic Disorders

The health problems caused by abnormalities in the genome of an individual are known as genetic disorders. These include Down syndrome, cystic fibrosis, Huntington's disease, sickle cell anemia, haemophilia, etc. The topics elaborated in this chapter will help in gaining a better perspective about these genetic disorders.

A genetic disorder is a disease that is caused by a change, or mutation, in an individual's DNA sequence.

- A genetic disorder is an illness caused by changes in a person's DNA.
- These mutations can be due to an error in DNA replication or due to environmental factors, such as cigarette smoke and exposure to radiation, which cause changes in the DNA sequence.
- The human genome is a complex set of instructions, like a recipe book, directing our growth and development.
- However, unlike a printed book, the human genome can change.
- These changes can affect the individual bases (A, C, G or T) or much larger chunks of DNA or even chromosomes.
- Our DNA provides the code for making proteins, the molecules that perform most of the functions in our body.
- However, when a section of our DNA is changed in some way, the protein it codes for is also affected and may no longer be able to carry out its normal function.
- Depending on where these mutations occur, they can have little or no effect, or may profoundly alter the biology of cells in our body, resulting in a genetic disorder.

Genetic disorders can be grouped into three main categories:

1. Single Gene Disorders: Disorders caused by defects in one particular gene, often with simple and predictable inheritance patterns.

- Dominant diseases: Single gene disorders that occur when an individual has one altered copy of the relevant gene and one healthy copy. For example, Huntington's disease.
- Recessive diseases: Single gene disorders that only occur when an individual has two altered versions of the relevant gene. For example, cystic fibrosis.
- X-linked disorders: Single gene disorders that reflect the presence of an altered gene on the X chromosome. X-linked disorders are more common in males because they only have one

X chromosome. As a consequence males only need one copy of the altered gene for symptoms to occur. For example, muscular dystrophy.

2. Chromosome Disorders: Disorders resulting from changes in the number or structure of the chromosomes. For example, Down's syndrome, which results from an extra chromosome 21 (trisomy 21: three copies of chromosome 21).

Down's syndrome is a chromosome disorder resulting from an extra chromosome 21.

3. Multifactorial disorders (complex diseases): Disorders caused by changes in multiple genes, often in a complex interaction with environmental and lifestyle factors such as diet or cigarette smoke. For example, cancer.

Down Syndrome

Down syndrome is a genetic disorder caused when abnormal cell division results in an extra full or partial copy of chromosome 21. This extra genetic material causes the developmental changes and physical features of Down syndrome.

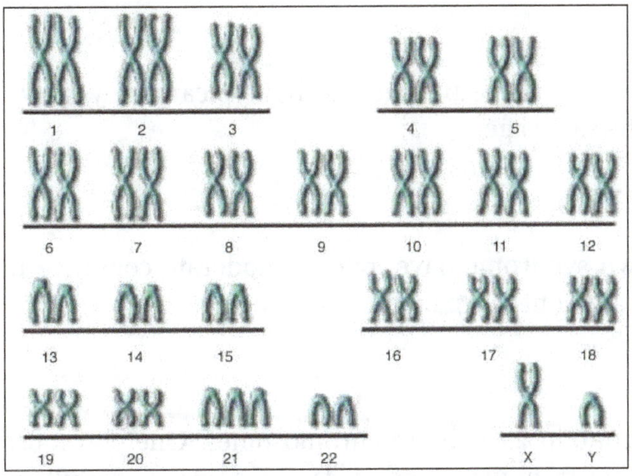

Down syndrome varies in severity among individuals, causing lifelong intellectual disability and developmental delays. It's the most common genetic chromosomal disorder and cause of learning disabilities in children. It also commonly causes other medical abnormalities, including heart and gastrointestinal disorders. Better understanding of Down syndrome and early interventions can greatly increase the quality of life for children and adults with this disorder and help them live fulfilling lives.

Symptoms

Each person with Down syndrome is an individual — intellectual and developmental problems may be mild, moderate or severe. Some people are healthy while others have significant health problems such as serious heart defects. Children and adults with Down syndrome have distinct facial features. Though not all people with Down syndrome have the same features, some of the more common features include:

- Flattened face,
- Small head,
- Short neck,
- Protruding tongue,
- Upward slanting eye lids (palpebral fissures),
- Unusually shaped or small ears,
- Poor muscle tone,
- Broad, short hands with a single crease in the palm,
- Relatively short fingers and small hands and feet,
- Excessive flexibility,
- Tiny white spots on the colored part (iris) of the eye called Brushfield's spots,
- Short height.

Infants with Down syndrome may be average size, but typically they grow slowly and remain shorter than other children the same age.

Intellectual Disabilities

Most children with Down syndrome have mild to moderate cognitive impairment. Language is delayed, and both short and long-term memory is affected.

Causes

Human cells normally contain 23 pairs of chromosomes. One chromosome in each pair comes from your father, the other from your mother. Down syndrome results when abnormal cell division involving chromosome 21 occurs. These cell division abnormalities result in an extra partial or full chromosome 21. This extra genetic material is responsible for the characteristic features and developmental problems of Down syndrome. Any one of three genetic variations can cause Down syndrome:

- Trisomy 21: About 95 percent of the time, Down syndrome is caused by trisomy 21 — the person has three copies of chromosome 21, instead of the usual two copies, in all cells. This is caused by abnormal cell division during the development of the sperm cell or the egg cell.
- Mosaic Down syndrome: In this rare form of Down syndrome, a person has only some cells with an extra copy of chromosome 21. This mosaic of normal and abnormal cells is caused by abnormal cell division after fertilization.
- Translocation Down syndrome: Down syndrome can also occur when a portion of chromosome 21 becomes attached (translocated) onto another chromosome, before or at conception. These children have the usual two copies of chromosome 21, but they also have additional genetic material from chromosome 21 attached to another chromosome.

There are no known behavioral or environmental factors that cause Down syndrome.

Is it Inherited?

Most of the time, Down syndrome isn't inherited. It's caused by a mistake in cell division during early development of the fetus. Translocation Down syndrome can be passed from parent to child. However, only about 3 to 4 percent of children with Down syndrome have translocation and only some of them inherited it from one of their parents. When balanced translocations are inherited, the mother or father has some rearranged genetic material from chromosome 21 on another chromosome, but no extra genetic material. This means he or she has no signs or symptoms of Down syndrome, but can pass an unbalanced translocation on to children, causing Down syndrome in the children.

Risk Factors

Some parents have a greater risk of having a baby with Down syndrome. Risk factors include:

- Advancing maternal age: A woman's chances of giving birth to a child with Down syndrome increase with age because older eggs have a greater risk of improper chromosome division. A woman's risk of conceiving a child with Down syndrome increases after 35 years of age. However, most children with Down syndrome are born to women under age 35 because younger women have far more babies.
- Being carriers of the genetic translocation for Down syndrome: Both men and women can pass the genetic translocation for Down syndrome on to their children.
- Having had one child with Down syndrome: Parents who have one child with Down syndrome and parents who have a translocation themselves are at an increased risk of having another child with Down syndrome. A genetic counselor can help parents assess the risk of having a second child with Down syndrome.

Complications

People with Down syndrome can have a variety of complications, some of which become more prominent as they get older. These complications can include:

- Heart defects: About half the children with Down syndrome are born with some type of congenital heart defect. These heart problems can be life-threatening and may require surgery in early infancy.
- Gastrointestinal (GI) defects: GI abnormalities occur in some children with Down syndrome and may include abnormalities of the intestines, esophagus, trachea and anus. The risk of developing digestive problems, such as GI blockage, heartburn (gastroesophageal reflux) or celiac disease, may be increased.
- Immune disorders: Because of abnormalities in their immune systems, people with Down syndrome are at increased risk of developing autoimmune disorders, some forms of cancer, and infectious diseases, such as pneumonia.
- Sleep apnea: Because of soft tissue and skeletal changes that lead to the obstruction of their airways, children and adults with Down syndrome are at greater risk of obstructive sleep apnea.
- Obesity: People with Down syndrome have a greater tendency to be obese compared with the general population.
- Spinal problems: Some people with Down syndrome may have a misalignment of the top two vertebrae in the neck (atlantoaxial instability). This condition puts them at risk of serious injury to the spinal cord from overextension of the neck.
- Leukemia: Young children with Down syndrome have an increased risk of leukemia.
- Dementia: People with Down syndrome have a greatly increased risk of dementia — signs and symptoms may begin around age 50. Having Down syndrome also increases the risk of developing Alzheimer's disease.
- Other problems: Down syndrome may also be associated with other health conditions, including endocrine problems, dental problems, seizures, ear infections, and hearing and vision problems.

For people with Down syndrome, getting routine medical care and treating issues when needed can help with maintaining a healthy lifestyle.

Life Expectancy

Life spans have increased dramatically for people with Down syndrome. Today, someone with Down syndrome can expect to live more than 60 years, depending on the severity of health problems.

Prevention

There's no way to prevent Down syndrome. If you're at high risk of having a child with Down syndrome or you already have one child with Down syndrome, you may want to consult a genetic counselor before becoming pregnant. A genetic counselor can help you understand your chances of having a child with Down syndrome. He or she can also explain the prenatal tests that are available and help explain the pros and cons of testing.

Cystic Fibrosis

Cystic Fibrosis is an inherited disorder that can lead to a variety of clinical manifestations. In an attempt to better understand the pathology of this disorder it is important to analyze the specific anatomical structures that will be impacted by this disease process.

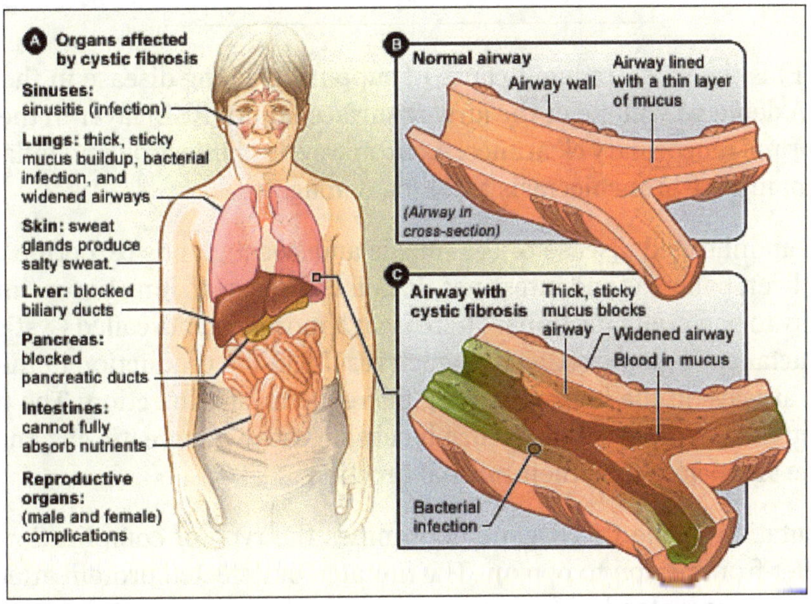

- Bronchioles: Obstructed secondary to the thickened mucus adhering to the walls of the airways.
- Small intestine: Distal obstruction (meconium ileus).
- Pancreatic duct: Obstruction.
- Bile ducts: Obstruction.
- Male reproductive organs: Absent or obstructed vas deferens.
- Female reproductive organs: Cervical canal obstructed by thickened mucus.
- Musculoskeletal system:
    - Inspiratory muscle atrophy.
    - Weakness/atrophy in anti- gravity muscles such as the gastrocnemius.
    - Kyphosis of the spine resulting in neck and back pain.

Mechanism of Injury/Pathological Process

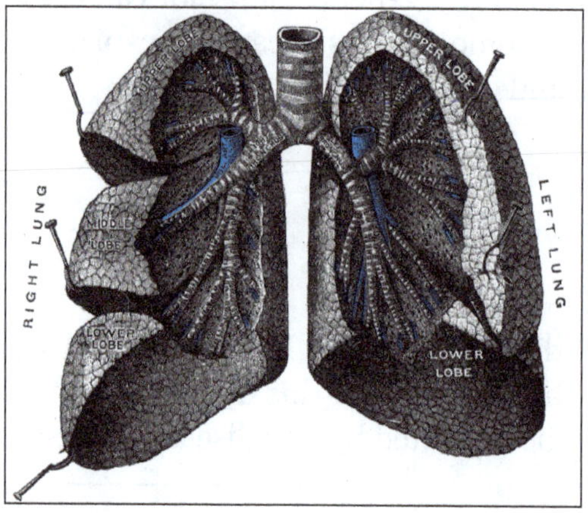

Cystic fibrosis (CF) is the most frequent cause of suppurative lung disease in the younger Caucasian population. A depleted volume of the airway surface liquid (ASL) layer in the respiratory system leads to abnormal mucociliary clearance. A chronic cycle of infection and inflammation results in progressive suppurative bronchiectasis and lung damage.

Cystic Fibrosis is an inherited disease of the mucus and sweat glands (exocrine glands) affecting mostly the lungs, liver, pancreas and intestines. It causes damage to lung tissue, inflammation, and acute susceptibility to bacterial infections. There is an abnormal gene, called Cystic Fibrosis Transmembrane Conductance Regulator (CFTR), which results in the production of thick, sticky mucus which blocks the airways in the lungs resulting in frequent lung infection. The thick mucus also blocks the ducts in the pancreas which in turn blocks digestive pancreatic enzymes from reaching the small intestine and performing their normal function.

The degree of mutation of the CFTR gene determines the type of complications that will arise and these will differ from person to person. The mutated or defected protein attaches to the outer membrane of cells in sweat glands, lungs, pancreas and other affected organs, and spans over the membrane which acts as a channel connecting the inner part of the cell cytoplasm to the surrounding fluid. This channel is important in our airways because it controls the movement of chloride from the inside to the outside of the cell. Chloride moves from the sweat into the cytoplasm but due to the mutation of the CFTR protein, the chloride is trapped inside the cells of the airway and outside the skin. Chloride is a negatively charged ion and sodium is a positively charged ion, so this leads to an electrical attraction, which results in the formation of salt. In CF patients a high amount of salt is lost in sweat, thus forming the basis of the sweat test. There are 3 theories:

- The lack of chloride exodus through the CFTR protein leads to the accumulation and an increase in the viscosity of the nutrient-rich mucus in the lungs leading to bacteria hiding from the body's immune system.
- There is a paradoxical increase in the sodium and chloride uptake due to the CFTR protein defect. This cause an increase in the water reabsorption which leads to thick and dehydrated mucus.
- This theory focuses on abnormal chloride movement out of the cell which leads to dehydrated mucus, pancreas, secretion, biliary secretion etc.

These theories propose that the major damage in patients with CF is due to the blockage of narrow passages of the affected organs due to thick secretions. The blockage within the lung airways leads to infections because of the accumulation of the enzymes and blockages of the passages.

Causes and Risk Factors

Cystic Fibrosis is an autosomal recessive disease caused by defects in the CFTR gene. This CFTR gene normally encodes for a protein that functions as a chloride channel and is regulated by cyclic AMP, but in cystic fibrosis, the CFTR gene mutates and results in abnormalities of cyclic AMP, which regulates the transport of chloride across the epithelial cells on mucosal surfaces. This defective gene causes the body to produce an abnormally thick and sticky fluid in the respiratory and GI tracts, the pancreas and in the sweat glands.

In order to develop CF, two CF genes must be inherited, one from the mother and the other from the father. If only one CF gene is inherited, then the person is called a carrier and will have no symptoms. When both parents are carriers, with each pregnancy there is a: 1 in 4 chance of having a child with cystic fibrosis, 1 in 2 chance of having a child who is a carrier or 1 in 4 chance of having an unaffected child. There are several different types of genetic mutation which are associated with different degrees of severity of the disease.

Signs and Symptoms

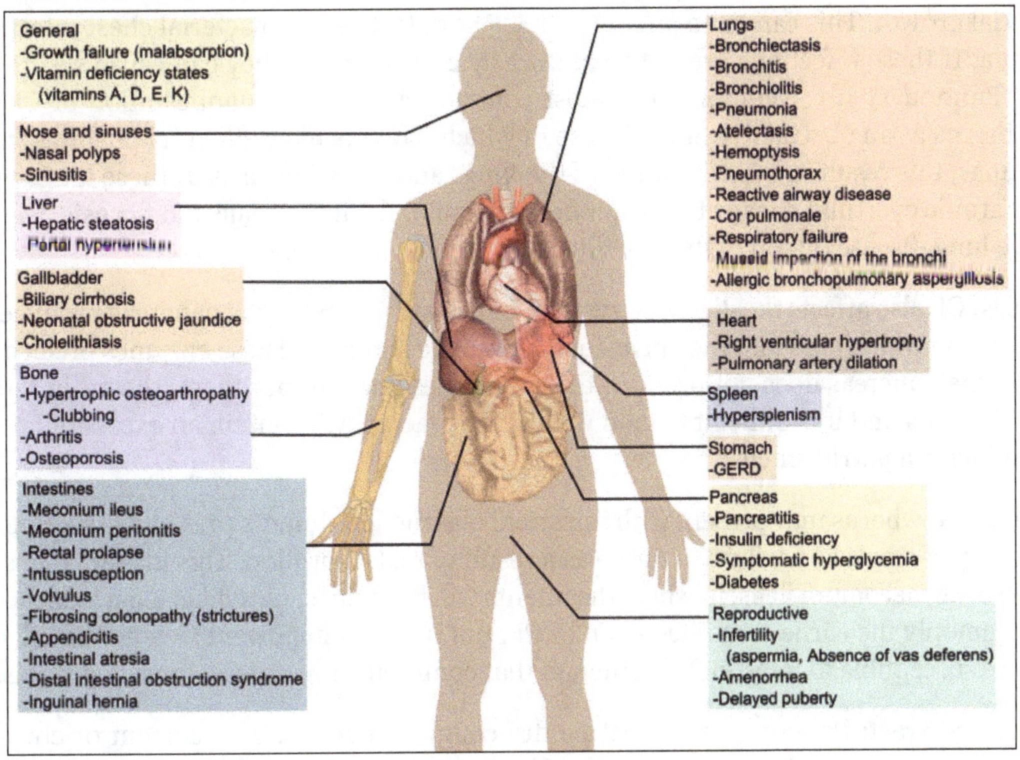

Manifestations of Cystic Fibrosis.

Cystic Fibrosis can be asymptomatic; this occurs when a person only inherits one defective gene from either of their parents and is a carrier. The pattern of development of CF and the severity of its symptoms varies among individuals. The disease is sometimes obvious soon after birth, but in some cases of CF, they are not detected for months (in infancy) or years (in childhood). Signs and symptoms will differ from person to person and these may include:

- Fatigue.
- Salty-tasting skin.
- Persistent cough with phlegm.
- Wheezing and shortness of breath.
- Lung infections.
- Poor growth and weight loss.
- Difficulty with bowel movements in the first 24/48 hours of life.

Organs Affected

CF is a "multi-system" disease, meaning that it affects many body organs. However, most of the symptoms involve the lungs and the digestive system.

1. Lungs: In a healthy person, there is a constant flow of mucus over the surfaces of the air passages in the lungs. This removes debris and bacteria. If you have CF, this mucus is excessively thick and sticky and cannot perform this role properly. The sticky mucus also provides an ideal environment for bacterial growth. This can put a person with CF at risk of getting bacterial chest infections and pneumonia. If these infections are not treated early and properly, they can cause severe complications. "Pulmonary involvement is the most common and severe manifestation of CF". Cystic Fibrosis begins as an obstructive lung disease in which patients have difficulty pushing the air out of their lungs, this result in hyperinflation of the lungs and bronchiectasis. If these lungs problems become chronic overtime it leads to further destruction of the lung tissue and fibrosis, leading to a restrictive lung disease on top of the already present obstructive lung disease.

2. Pancreas: CF also affects the digestive system. In a healthy person, the pancreas produces chemicals (enzymes) which pass into the gut as food leaves the stomach. These enzymes break down the fat. In CF, the pancreas does not produce these enzymes and without them, the fat in food is not properly digested and it is difficult to gain weight. The faeces will contain an excess of fat and will be oily and have a putrid smell.

3. Intestines: Newborns may present with intestinal obstruction. Infants present with an increase in the frequency of stools and failure to thrive as a result of malabsorption. This intestinal obstruction is referred to as meconium ileus, in which the infants will have thick, puttylike, tenacious meconium. This is commonly the earliest manifestation of CF, but it is also important to note that all children with CF are susceptible to intestinal obstruction that commonly presents as impacted stools.

4. Respiratory Tract: Patients present with a dry cough, which can be recurrent or chronic, and can also produce sputum. The patient may suffer from shortness of breath and chest pain. There may also be recurrent wheezing and pneumonia, pneumothorax and haemoptysis. In babies and infants, the symptoms of CF are:

- Persistent diarrhea,

- Bulky, foul smelling and greasy stools,
- Pale stools,
- Frequent wheezing or pneumonia,
- Chronic cough with thick mucus,
- Salty-tasting skin,
- Poor growth,
- Blockage of the intestine (called meconium ileus),
- Abdominal swelling,
- Gassiness,
- Vomiting,
- Dehydration.

In children, the symptoms include:

- Frequent respiratory infections,
- Fever,
- Cough,
- Difficulty in breathing,
- Abdominal pain and discomfort,
- Flatulence,
- Fast respiration,
- Flaring of the nostrils,
- Poor appetite,
- Malnutrition,
- Poor growth,
- A barrel-chested appearance.

5. Genitourinary System: Infertility in men and women is a common clinical manifestation of CF. Men have decreased fertility secondary to the absence or obstruction of the vas deferens and women have decreased fertility secondary to the thickened mucus in the cervical canal.

6. Musculoskeletal System: It is important for Physical Therapist to be aware of the clinical manifestations in the musculoskeletal system. Physical Therapists should be aware that these patients are likely to have decreased bone mineral density, which is going impact the spine. Thesc patients will often have increased Thoracic Hyperkyphosis secondary to becoming barrel chested. Other manifestations may include muscle atrophy and myalgia. Adults who have been dealing with

Cystic Fibrosis long-term may present with osteoarthropathy, Rheumatoid arthritis, and/or osteoporosis. Other medical problems may arise as a result of CF, such as:

- Sinusitis (inflammation of the nasal sinuses),
- Nasal polyps (fleshy growths inside the nose),
- Clubbing (rounding and enlargement of fingers and toes),
- Pneumothorax (rupture of lung tissue and trapping of air between the lung and chest wall),
- Coughing up blood,
- Enlargement of the right side of the heart (called cor pulmonale),
- Protrusion of the rectum through the anus (called rectal prolapse),
- Liver, pancreatic and gallbladder problems,
- Delayed puberty.

Huntington's Disease

Huntington's disease (HD) is a brain disease that is passed down in families from generation to generation. It is caused by a mistake in the DNA instructions that build our bodies and keep them running. DNA is made up of thousands of genes, and people with HD have a small error in one gene, called huntingtin. Over time this error causes damage to the brain and leads to HD symptoms. HD causes deterioration in a person's physical, mental, and emotional abilities, usually during their prime working years, and currently has no cure. Most people start developing symptoms during adulthood, between the ages of 30 to 50, but HD can also occur in children and young adults (known as juvenile HD or JHD). HD is known as a family disease because every child of a parent with HD has a 50/50 chance of inheriting the faulty gene. Today, there are approximately 41,000 symptomatic Americans and more than 200,000 at-risk of inheriting the disease.

Symptoms of Huntington's Disease

The symptoms of HD can vary a lot from person to person, but they usually include:

- Personality changes, mood swings & depression,
- Forgetfulness & impaired judgment,
- Unsteady gait & involuntary movements (chorea),
- Slurred speech, difficulty in swallowing & significant weight loss.

Most people with HD experience problems with thinking, behavior, and movements. Symptoms usually worsen over the course of 10 to 25 years and affect the ability to reason, walk, and talk. Early on, a person with HD or their friends and family may notice difficulties with planning, remembering, and staying on task. They may develop mood changes like depression, anxiety, irritability, and anger. Most people with HD become "fidgety" and develop movements of the face and limbs known as chorea, which they are not able to control.

Because of the uncontrolled movements (chorea), a person with HD may lose a lot of weight without intending to, and may have trouble walking, balancing, and moving around safely. They will eventually lose the ability to work, drive, and manage tasks at home, and may qualify for disability benefits. Over time, the individual will develop difficulty with speaking and swallowing, and their movements will become slow and stiff. People with advanced HD need full-time care to help with their day-to-day activities, and they ultimately succumb to pneumonia, heart failure or other complications. The symptoms of HD are sometimes described as having ALS, Parkinson's and Alzheimer's – simultaneously.

The Huntingtin Gene and Protein

The DNA error that causes HD is found in a gene called huntingtin. This gene was discovered in 1993. Everyone has the huntingtin gene, but only those that inherit the mistake, known as the HD mutation, will develop HD and risk passing it on to their children. Genes are made up of the nucleotide "letters" A,G,C, and T, which form a code that is read in groups of three. HD is caused by a stretch of the letters C-A-G in the huntingtin gene which repeat over and over, too many times CAGCAGCAGCAGCAG. This is known as a CAG repeat expansion. In the huntingtin gene, most people have around 20 CAG repeats, but people with HD have around 40 or more. Every person who has this CAG repeat expansion in the HD gene will eventually develop the disease, and each of their children has a 50% chance of developing HD.

Our genes are like an instruction manual for making proteins, the machines that run everything in our bodies. The huntingtin gene (DNA) contains instructions that are copied into a biological message (RNA) which makes the huntingtin protein. The huntingtin protein is very large and seems to have many functions, especially as the brain is developing before birth, but it is not fully understood. We know that the extra CAG repeats in people with HD cause the huntingtin protein to be extra-long and difficult to maintain, which makes it difficult for it to do its job. Over many years, this "mutant" huntingtin protein forms clumps in brain cells, and causes them to become damaged and die. The most vulnerable part of the brain in HD is called the striatum, and it controls movement, mood, and memory. Damage to the striatum over time is what causes the symptoms of HD.

Treating Huntington's Disease

There is currently no cure or treatment which can halt, slow or reverse the progression of the disease. However, there are many treatments and interventions that can help to manage HD symptoms. A neurologist, psychiatrist, or nurse with expertise in HD may prescribe medications to ease anxiety and depression, help with troublesome behaviors, and calm uncontrolled movements. A psychologist or social worker can provide individual or group counseling. Physical and occupational therapists can work with patients and families to develop strength, move safely, and adjust the home environment and activities as needed. Speech language pathologists and nutritionists can help with communication, eating and swallowing safely, and combating weight loss. Clinician researchers may suggest participation in HD clinical trials.

Social and community support is an important part of HD care. Family, friends, loved ones, and companions often assume many of the HD person's former responsibilities and help with daily activities and care routines when they can no longer do so themselves. Caregivers and kids may also need support for the challenges and stresses that come with HD.

Duchenne Muscular Dystrophy

Muscular dystrophies are characterized by specific abnormalities (e.g. variation of muscle fiber size, muscle fiber necrosis, scar tissue formation and inflammation) in muscle biopsy from the patients. Approximately 30 different genetic conditions make up the muscular dystrophies. DMD is classified as a dystrophinopathy. The dystrophinopathies are a spectrum of muscle diseases, each caused by alterations in the dystrophin gene. The most severe end of the spectrum is known as Duchenne muscular dystrophy lacking completely dystrophin protein. Decreased or truncated dystrophin protein is associated with less severe form is Becker muscular dystrophy. The clinical hallmarks of DMD include weakness and wasting of various voluntary muscles of the body. In most advanced stages of the disease, the heart and gut muscles will be affected.

Signs and Symptoms

DMD usually becomes apparent early during childhood. Affected children develop weakness and wasting (atrophy) of the muscles closest to the trunk (proximal muscles) such as those of the upper legs and pelvic area and upper arms and shoulder area. However, a few other muscles appear disproportionally bulky. As the disease progresses, muscle weakness and atrophy spread to affect the lower legs, forearms, neck and trunk. The rate of progression is quite similar from person to person but individual variation may happen.

In children with DMD, initial findings may include delays in reaching developmental milestones such as sitting or standing without assistance; toe walking; an unusual, waddling manner of walking (gait); difficulty climbing stairs or rising from a sitting position (Gower's sign); and repeated falling. Toddlers and young children may seem awkward and clumsy and may exhibit abnormal enlargement of the calves due to scarring of muscles (pseudohypertrophy). Parents may be falsely encouraged by an apparent improvement between the ages of 3 and 5, but this may be due to natural growth and development. As the disease progresses, additional abnormalities may develop such as progressive curvature of the spine (scoliosis or lordosis), wasting of thigh and pectoral muscles, and abnormal fixation of certain joints (contractures). A contracture occurs when thickening and shortening of tissue such as muscle fibers causes deformity and restricts movement of affected areas, especially the joints. Without physical therapy treatment, leg braces may be needed by age 8-9 to assist affected individuals to walk. By approximately ages 10 to 12, most affected individuals require a wheelchair.

Children with DMD have reduced bone density and an increased risk of developing fractures of certain bones, such as hips and spine. Many affected individuals will display mild to moderate degrees of non-progressive intellectual impairment and learning disabilities. By the late teens, DMD may also be characterized by additional potentially life-threatening complications including weakness and deterioration of the heart muscle (cardiomyopathy). Cardiomyopathy can result in impairment in the ability of the heart to pump blood, irregular heartbeats (arrhythmias), and heart failure. Another serious complication associated with DMD is weakness and deterioration of muscles in the rib cage. This can result in an increased susceptibility to respiratory infections (e.g., pneumonia), difficulty coughing, and, ultimately, respiratory failure.

Involvement of muscles within the gastrointestinal tract may result in dysmotility, a condition in which the passage of food through the digestive tract usually because of slow and uncoordinated movements of the muscles of the digestive tract. Gastrointestinal dysmotility may result in constipation and diarrhea. One third of patients with DMD may have various degree of cognitive impairment including learning disability, attention deficit and autistic spectrum disorder.

Causes

DMD is inherited as an X-linked disease. X-linked genetic disorders are conditions caused by an abnormal gene on the X chromosome and manifest mostly in males. Females that have a defective gene present on one of their X chromosomes are carriers for that disorder. Carrier females usually do not display symptoms because females have two X chromosomes and only one carries the defective gene. Males have one X chromosome that is inherited from their mother and if a male inherits an X chromosome that contains a defective gene he will develop the disease.

Female carriers of an X-linked disorder have a 25% chance with each pregnancy to have a carrier daughter like themselves, a 25% chance to have a non-carrier daughter, a 25% chance to have a son affected with the disease and a 25% chance to have an unaffected son. If a male with an X-linked disorder is able to reproduce, he will pass the defective gene to all of his daughters who will be carriers. A male cannot pass an X-linked gene to his sons because males always pass their Y chromosome instead of their X chromosome to male offspring. Some females who inherit a single copy of the disease gene for DMD (gene carriers or heterozygotes) may exhibit some of the symptoms associated with the disease such as weakness of certain muscles, especially those of the arms, legs, and back Carrier females who develop symptoms of DMD are also at risk for developing heart abnormalities, which may present as exercise intolerance or shortness of breath. If left untreated, heart abnormalities can cause life-threatening complications in such affected females.

DMD is caused by mutations of the DMD gene located on the short arm (p) of the X chromosome (Xp21.2). Chromosomes, which are present in the nucleus of human cells, carry the genetic information for each individual. Human body cells normally have 46 chromosomes. Pairs of human chromosomes are numbered from 1 through 22 and the sex chromosomes are designated X and Y. Males have one X and one Y chromosome and females have two X chromosomes. Each chromosome has a short arm designated "p" and a long arm designated "q". Chromosomes are further sub-divided into many bands that are numbered. For example, "chromosome Xp21.2" refers to band 21.2 on the short arm of the X chromosome. The numbered bands specify the location of the thousands of genes that are present on each chromosome.

The DMD gene regulates (encodes for) the production of dystrophin, a protein that appears to play an essential role in maintaining the integrity of cell membrane in skeletal (voluntary) and cardiac muscle cells. Dystrophin is found attached to the inner side of the membrane that surrounds muscle fibers. Mutation of the DMD gene will result in absence of the dystrophin protein, leading to degeneration of muscle fibers. The body can replace (regenerate) some muscle fibers, but over time more and more muscle fiber is lost. Such degeneration leads to the symptoms and findings associated with DMD. In Becker muscular dystrophy, a related disorder, dystrophin is present, but it is truncated or only present in insufficient levels to properly perform its functions. Although most boys with DMD inherit the abnormal gene from their mothers, some may develop the diseases as the result of a spontaneous mutation of the dystrophin gene that occurs randomly for unknown reasons (de novo or sporadic cases).

Affected Populations

DMD is the most common childhood onset form of muscular dystrophy and affects males almost exclusively. The birth prevalence is estimated to be 1 in every 3,500 live male births. Age of onset is usually between 3 and 5 years of age. The muscular dystrophies as a whole are estimated to affect 250,000 individuals in the United States.

Related Disorders

Symptoms of the following disorders can be similar to those of DMD. Comparisons may be useful for a differential diagnosis. Becker muscular dystrophy is in the category of inherited muscle wasting diseases caused by gene abnormalities (mutations) that result in deficient or abnormal production of the dystrophin protein (dystrophinopathies). The abnormal gene is the same as for DMD and is located on the X chromosome. Becker muscular dystrophy also follows X-linked inheritance so it mostly affects males, but some female carriers are affected. Becker muscular dystrophy usually begins in the teens or early twenties, but can begin as late as the sixties and symptoms vary greatly between affected individuals. Muscle weakness and deterioration progress slowly but usually results in the need for a wheelchair. Muscles of the heart deteriorate (cardiomyopathy) in some affected individuals more seriously than the skeletal muscles of the body, and this process can become life threatening potentially causing heart failure. Learning disabilities involving visual abilities may be present but rarely. In Becker muscular dystrophy, dystrophin levels are reduced where in DMD they are absent or nearly absent. Consequently, the symptoms of these two disorders are similar, but most cases of Becker muscular dystrophy are less severe.

Emery-Dreifuss muscular dystrophy (EDMD) is a rare, often slowly progressive genetic disorder affecting the muscles of the arms, legs, face, neck, spine and heart. The disorder consists of the clinical triad of weakness and degeneration (atrophy) of certain muscles, joints that are fixed in a flexed or extended position (contractures), and abnormalities affecting the heart (cardiomyopathy) in mainly adults. Major symptoms may include muscle wasting and weakness particularly in arms and lower legs (humeroperoneal regions) and contractures of the elbows, Achilles tendons, and upper back muscles. In some patients, additional abnormalities may be present. In most cases, EDMD is inherited as an X-linked or autosomal dominant trait. In extremely rare cases, autosomal recessive inheritance has been reported. Although EDMD has different modes of inheritance, the symptoms are nearly the same.

Limb-girdle muscular dystrophy (LGMD) is a general term for a group of rare progressive genetic disorders that are characterized by wasting (atrophy) and weakness of the voluntary muscles of the hip and shoulder areas (limb-girdle area). Muscle weakness and atrophy are progressive and may spread to affect other muscles of the body. Approximately 15 different subtypes have been identified based upon abnormal changes (mutations) of certain genes. The age of onset, severity, and progression of symptoms of these subtypes varies greatly even among individuals in the same family. Some individuals may have a mild, slowly progressive form of the disease; other may have a rapidly progressive form that may cause severe disability. The various forms of LGMDs can now be distinguished by genetic and/or protein analysis. The various forms of LGMD may be inherited as an autosomal dominant or recessive trait. Autosomal dominant LGMD is known as LGMD1 and has five subtypes (LGMDA-E). Autosomal recessive LGMD is known as LGMD2 and has 10 subtypes (LGMDA-J).

Spinal muscular atrophy (SMA) that is caused by a deletion of the SMN gene on chromosome 5 is an inherited progressive neuromuscular disorder characterized by degeneration of groups of nerve cells (lower motor neurons) within the lowest region of the brain (lower brainstem) and certain motor neurons in the spinal cord (anterior horn cells). Motor neurons are nerve cells that transmit nerve impulses from the spinal cord or brain (central nervous system) to muscle or glandular tissue. Typical symptoms are a slowly progressive muscle weakness and muscle wasting (atrophy). Affected individuals have poor muscle tone, muscle weakness on both sides of the body without, or with minimal, involvement of the face muscles, twitching tongue and a lack of deep tendon reflexes. SMA is divided into subtypes based on age of onset of symptoms and maximum function achieved.

Sickle Cell Anemia

Sickle cell anemia is a hereditary disease that destroys red blood cells by causing them to take on a rigid "sickle" shape. The disease is characterized by many of the symptoms of chronic anemia (fatigue, pale skin, and shortness of breath) as well as susceptibility to infection, jaundice and other eye problems, delayed growth, and episodic crises of severe pain in the abdomen, bones, or muscles. Sickle cell anemia occurs mainly in persons of African descent. The disease also occurs in persons of Middle Eastern, Mediterranean, and Indian descent.

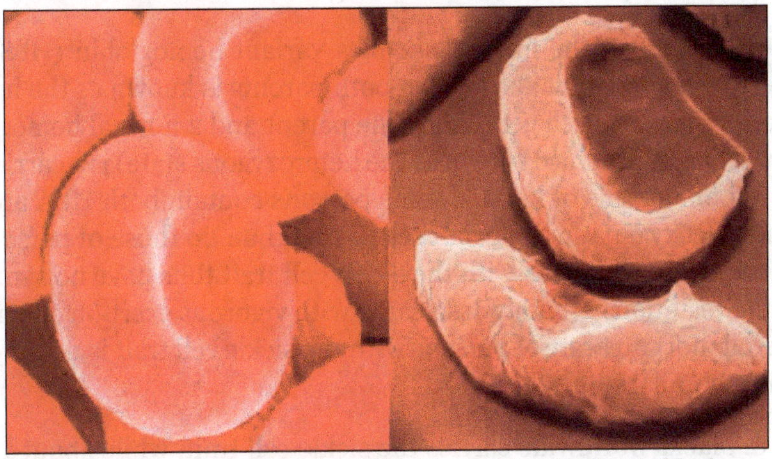

Blood cells in sickle cell anemia compared with healthy red blood cells. Healthy human red blood cells (left) compared with red blood cells from a person with sickle cell anemia (right).

Sickle cell anemia is caused by a variant type of hemoglobin, the protein in red blood cells that carries oxygen to the tissues of the body, called hemoglobin S (HbS). HbS is sensitive to deficiency of oxygen. When the carrier red blood cells release their oxygen to the tissues and the oxygen concentration within those cells is reduced, HbS, in contrast to normal hemoglobin (HbA), becomes stacked within the red cells in filaments that twist into helical rods. These rods then cluster into parallel bundles that distort and elongate the cells, causing them to become rigid and assume a sickle shape. This phenomenon is to some extent reversible after the cells become oxygenated once more, but repeated sickling ultimately results in irreversible distortion of the red cells. The sickle-shaped cells become clogged in small blood vessels, causing obstruction of the microcirculation, which in turn results in damage to and destruction of various tissues.

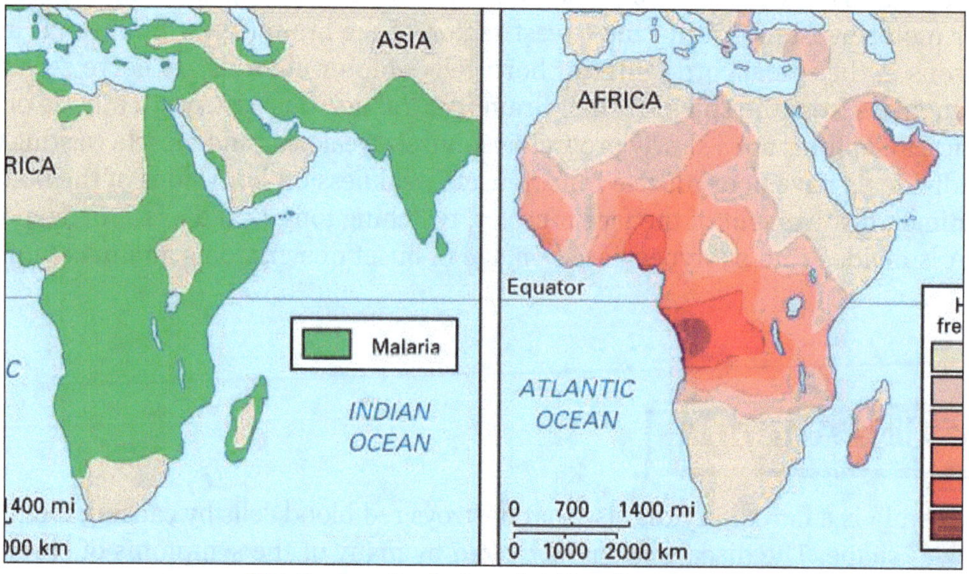

Distribution of malaria and sickle cell anemia.

The distribution of malaria and the distribution of sickle cell anemia overlap in areas of Africa, southern Asia, and the Mediterranean. The persistence of the HbS gene, which causes sickle cell anemia, has been explained by the fact that heterozygous persons are resistant to malaria.

Sickle cell anemia is caused by the inheritance of a variant hemoglobin (HbS) gene from both parents. (This inheritance of variant genes from both parents is known as the homozygous state.) A person who inherits the sickle cell gene from one parent and a normal hemoglobin gene (HbA) from the other parent (an inheritance known as the heterozygous state) is a carrier of the sickle cell trait. Because the red blood cells of heterozygous persons contain both HbA and HbS, such cells require much greater deoxygenation to produce sickling than do those of persons with sickle cell anemia. The great majority of persons with the sickle cell trait thus have no symptoms of disease, although certain manifestations—mainly associated with vigorous exertion at high altitudes—have been seen. The overall mortality rate of persons with the sickle cell trait is no different from that of a normal comparable population.

An estimated 1 in 12 blacks worldwide carries the sickle cell trait, while about 1 in 400 has sickle cell anemia. If both parents have the sickle cell trait, the chances are 1 in 4 that a child born to them will develop sickle cell anemia. However, through amniocentesis (analysis of amniotic fluid surrounding a fetus), a testing procedure done in the early stages of pregnancy, it is possible to detect sickle cell anemia in the fetus.

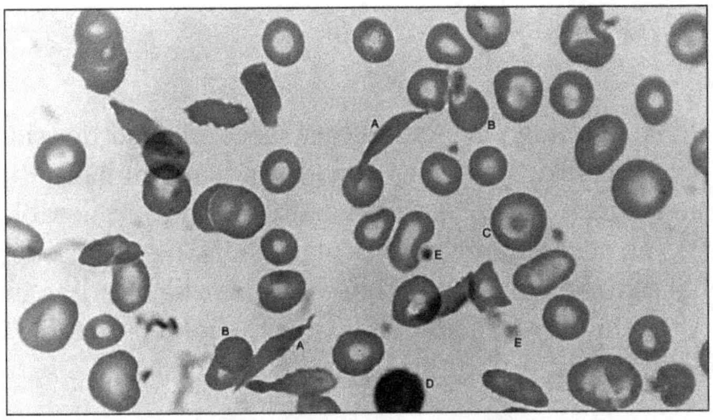

Blood smear; sickle cell anemia.

Blood smear in which the red cells show variation in size and shape typical of sickle cell anemia. (a) Long, thin, deeply stained cells with pointed ends are irreversibly sickled. (b) Small, round, dense cells are hyperchromic because a part of the membrane is lost during sickling. (c) Target cell with a concentration of hemoglobin on its centre. (d) Lymphocyte. (e) Platelets.

The HbS gene is distributed geographically in a broad equatorial belt in Africa and also is found, though less often, in other parts of the continent and in the Americas. The persistence of HbS has been explained by the fact that heterozygous persons are resistant to malaria. When the red cells of a person with the sickle cell trait are invaded by the malarial parasite, the red cells adhere to blood vessel walls, become deoxygenated, assume the sickle shape, and then are destroyed, the parasite being destroyed with them.

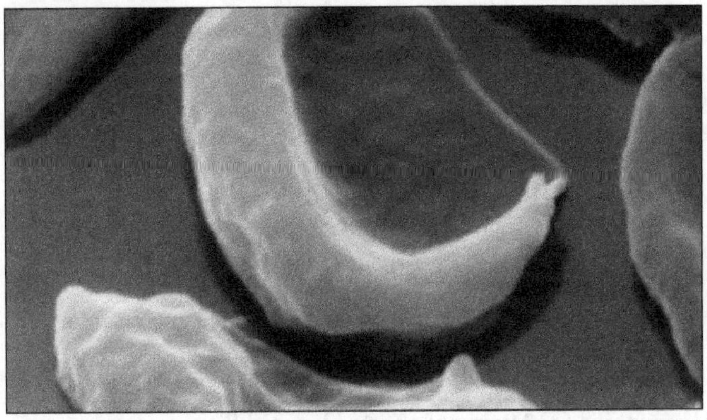

Photomicrograph of red blood cells, showing abnormal shape characteristic of sickle cell anemia.

There is no cure for sickle cell anemia; most care is devoted to alleviating symptoms. Infants and young children with the disease are given regular daily doses of penicillin to prevent serious infection. In some cases blood transfusions are given regularly to prevent organ damage and stroke and to relieve the worst symptoms of red blood cell loss. In severe cases bone marrow transplantation has been of some benefit. The drug hydroxyurea reduces the principal symptoms of sickle cell anemia. Hydroxyurea apparently activates a gene that triggers the body's production of fetal hemoglobin. This type of hemoglobin, which is ordinarily produced in large amounts only by infants shortly before and after birth, does not sickle. Hydroxyurea therapy increases the proportion of fetal hemoglobin in the bloodstream of adult patients from 1 to about 20 percent, a proportion high enough to lessen markedly the circulatory problems that arise during crises.

Haemophilia

Hemophilia, also spelled haemophilia is a hereditary bleeding disorder caused by a deficiency of a substance necessary for blood clotting (coagulation). In hemophilia A, the missing substance is factor VIII. The increased tendency to bleeding usually becomes noticeable early in life and may lead to severe anemia or even death. Large bruises of the skin and soft tissue are often seen, usually following injury so trivial as to be unnoticed. There may also be bleeding in the mouth, nose, and gastrointestinal tract. After childhood, hemorrhages in the joints—notably the knees, ankles, and elbows—are frequent, resulting in swelling and impaired function.

The transmission of this condition is characteristically sex-linked, being expressed almost exclusively in males but transmitted solely by females; sons of a male with hemophilia are normal, but daughters, although outwardly normal, may transmit the trait as an overt defect to half their sons and as a recessive or hidden trait to half their daughters, as shown in the chart. The existence of hemophilia in certain royal families of Europe, particularly descendants of Great Britain's Queen Victoria, is well known.

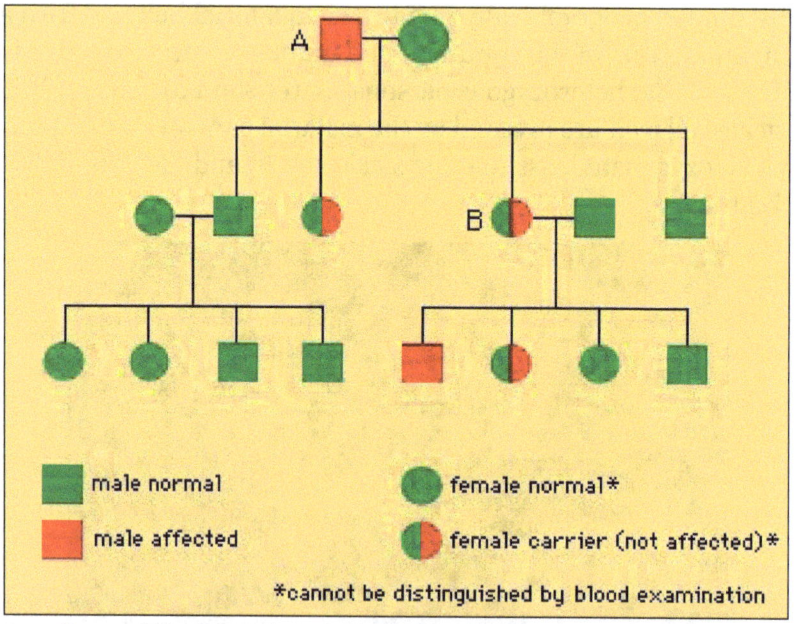

Transmission of hemophilia (A) Mating of affected hemophilic man and normal woman—all sons normal, all daughters carriers. (B) Mating of carrier woman and normal man—half of sons normal and half affected; half of daughters carriers, half normal.

Hemophilia may also be attributed to a deficiency of factor IX (hemophilia B) or of factor XI (hemophilia C); hemophilia B (also called Christmas disease), like hemophilia A, is sex-linked and occurs almost only in males, whereas hemophilia C may be transmitted by both males and females and is found in both sexes. Persons with hemophilia are ordinarily advised to avoid activities that might expose them to bodily injury. The management of bleeding episodes includes infusions of clotting factor, which is derived from human blood or by recombinant DNA technology. The drug desmopressin (DDAVP) is useful in treating milder forms of hemophilia A. Gene therapy has been shown to be effective in correcting factor IX deficiency associated with hemophilia B.

Thalassemia

Thalassemia is a genetic blood disorder. People with Thalassemia disease are not able to make enough hemoglobin, which causes severe anemia. Hemoglobin is found in red blood cells and carries oxygen to all parts of the body. When there is not enough hemoglobin in the red blood cells, oxygen cannot get to all parts of the body. Organs then become starved for oxygen and are unable to function properly. There are two primary types of Thalassemia disease: Alpha Thalassemia disease and Beta Thalassemia disease. Beta Thalassemia Major (also called Cooley's Anemia) is a serious illness. Symptoms appear in the first two years of life and include paleness of the skin, poor appetite, irritability, and failure to grow. Proper treatment includes routine blood transfusions and other therapies. There are two main types of Alpha Thalassemia disease. Alpha Thalassemia Major is a very serious disease in which severe anemia begins even before birth. Pregnant women carrying affected fetuses are themselves at risk for serious pregnancy and delivery complications. Another type of Alpha Thalassemia is Hemoglobin H disease. There are varying degrees of Hemoglobin H disease.

Thalassemia is a complex group of diseases that are relatively rare in the United States but common in Mediterranean regions and South and Southeast Asia. Worldwide, there are 350,000 births per year with serious hemoglobinopathies (blood disorders). In the United States, as a consequence of immigration patterns, occurrence of thalassemia disorders is increasing. The thalassemias are a diverse group of genetic blood diseases characterized by absent or decreased production of normal hemoglobin, resulting in a microcytic anemia of varying degree. The thalassemias have a distribution concomitant with areas where P. falciparum malaria is common. The alpha thalassemias are concentrated in Southeast Asia, Malaysia, and southern China. The beta thalassemias are seen primarily in the areas surrounding Mediterranean Sea, Africa and Southeast Asia. Due to global migration patterns, there has been an increase in the incidence of thalassemia in North America in the last ten years, primarily due to immigration from Southeast Asia.

In the normal adult, hemoglobin A, which is composed of two alpha and two beta globins (A^2B^2), is the most prevalent, comprising about 95% of all hemoglobin. Two minor hemoglobins also occur: hemoglobin A_2, composed of two alpha and two delta globins (α^2 δ^2) comprises 2-3.5% of hemoglobin, while hemoglobin F, composed of two alpha and two gamma globins (α^2 γ^2) comprises less than 2% of hemoglobin. Hemoglobin F, or fetal hemoglobin, is produced by the fetus in utereo and until about 48 weeks after birth. Hemoglobin F has a high oxygen-affinity in order to attract oxygen from maternal blood and deliver it to the fetus. After birth, the production of adult hemoglobin rapidly increases and fetal hemoglobin production drops off.

The genes controlling globin production are on chromosome 16 (alpha globin genes: "α"), and chromosome 11 (beta: "β", gamma: "γ", and delta: "δ" genes). As seen in the diagram, the alpha globin molecule concentration is rather stable in fetal and adult life, because it is needed for both fetal and adult hemoglobin production. The beta globin appears early in fetal life at low levels and begins to rapidly increase after 30 weeks gestational age, reaching a maximum about 30 weeks postnatally. The gamma globin molecule reaches a high level early in fetal life at about 6 weeks and begins to decline about 30 weeks gestational age, reaching a low level about 48 weeks postgestational age. The delta globin appears at a low level at about 30 weeks gestational age and maintains a low profile throughout life.

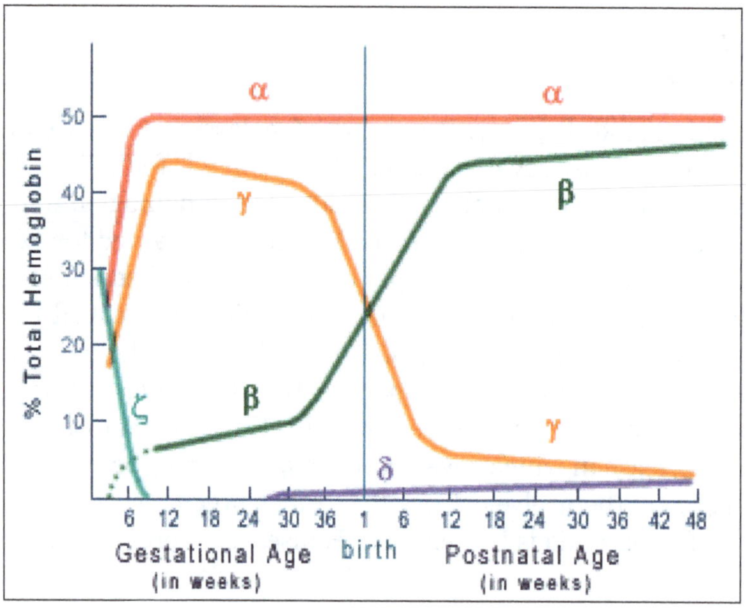

In the thalassemia patient, a mutation or deletion of the genes that control globin production occurs. This leads to a decreased production of the corresponding globin chains and an abnormal hemoglobin ratio (α: non-α). This abnormal ratio leads to decreased synthesis of hemoglobin and the expression of thalassemia. The globin that is produced in normal amounts winds up in excess and forms red cell aggregates or inclusions. These aggregates become oxidized and damage the cell membrane, leading either to hemolysis, ineffective erythropoiesis, or both. The quantity and properties of these globin chain aggregates determine the characterstics and severity of the thalassemia. Beta thalassemia results in an excess of alpha globins, which leads to the formation of alpha globin tetramers (α^4) that accumulate in the erythroblast (immature red blood cell). These aggregates are very insoluble and precipitation interferes with erythropoiesis, cell maturation and cell membrane function, leading to ineffective erythropoiesis and anemia.

Alpha thalassemia results in an excess of beta globins, which leads to the formation of beta globin tetramers (β^4) called hemoglobin H. These tetramers are more stable and soluble, but under special circumstances can lead to hemolysis, generally shortening the life span of the red cell. Conditions of oxidant stress cause Hgb H to precipitate, interfering with membrane function and leading to red cell breakage. Hemoglobin H-Constant Spring disease is a more severe form of this hemolytic disorder. The most severe thalassemia is alpha thalassemia major, in which a fetus produces no alpha globins, which is generally incompatible with life.

Fragile X Syndrome

The Fragile X Syndrome (FXS) is a genetic disease inherited through the X chromosome, which was described for the first time in 1943 by Martin and Bell. It is actually considered the most common inherited cause of intellectual disability and the second most prevalent cause after Down syndrome. Most cases of Down syndrome are de novo but FXS is always inherited with many individuals in the family tree, either affected or a carrier of FXS. Affected men have a classic phenotype characterized by long face, large and protruding ears and macroorchidis . The Fragile X Syndrome is caused by an abnormal expansion in the number of the trinucleotide CGG repeats located in the 5' UTR in the fragile X mental retardation 1 gene (*FMR1*) at Xq27.3. It is a dynamic mutation with expansion of the CGG repeat in each generation moving from the premutation range of 55 to 200 repeats and expanding to a full mutation when pass on by a women to her children.

Patients affected with FXS have more than 200 repeats of the CGG trinucleotide. On the other hand, premutation carriers (55 to 200 repeats) although are not affected with the classic FXS phenotype, can have other medical, psychiatric and neurological problems. In the last 15 years multiple advances have been made in the description of genetic characteristics, function of the protein encoded by the *FMR1* gene (FMRP), pharmacological management and the description, in carriers of the premutation, of the Fragile X associated Tremor/Ataxia Syndrome (FXTAS) and fragile X-associated primary ovarian insufficiency (FXPOI).

Genomics

The Fragile X Syndrome is caused by an alteration in the *FMR1* gene, with locus Xq27.3. This gene harbors a CGG repeat within the 5' Untranslated Region. Depending on the number of repetitions, 4 types of alleles are defined with different clinical manifestations, Normal alleles, up to 44 CGG repeats; premutation (PM) alleles, between 55 and 200 and full mutation alleles (FM) with more than 200 repeats. The fourth type of allele is named "grey zone" or intermediate allele and contains between 45 and 54 repeats and it has been proposed as a precursor for PM alleles.

The silencing of the *FMR1* gene is the result of a series of complex epigenetic modifications following the expansion of the trinucleotide repeat. The FM alleles undergo a methylation process in the CpG Island within the gene promoter and in the CGG repeats. In male patients with the FM allele, every cytosine within the CpG Island is methylated, contrarily to healthy individuals who lack of any methylation. A recently discovered boundary sequence in the *FMR1* gene found 650 to 800 nucleotides upstream from the repeated region that suffers methylation. This boundary sequence, because of its chromatin interaction, limits the hypermethylated region of the genome, protecting the promoter of *FMR1* from possible methylation. In individuals with FXS the methylated boundary sequence is lost allowing the expansion of the methylation up to the promoter of the *FMR1* gene. This finding strongly suggests that changes in the sequence of nucleotides and conformational structure of the chromatin of the boundary sequences would favor the epigenetic changes that would induce *FMR1* silencing and ultimately preventing FMRP production.

The Fragile X Syndrome is usually caused by the methylation and gene silencing associated with the full mutation although deletions of the coding region of the gene can also lead to absence of FMRP. In addition, point mutations or reading frame shifts can also occur leading to a functional deficit of the protein FMRP and the consequent phenotype; however, these genomic changes account only for less than 1% of all FXS cases described. The clinical involvement in those with a premutation is caused by a different mechanism than FXS, which is elevated *FMR1* mRNA levels leading to RNA toxicity. 2-8 folds normal levels of mRNA are observed in premutation carriers and the excess such mRNA lead to the sequestration of important proteins for neuronal function by the hairpin structures that form in the CGG repeats. RNA toxicity causes the neurons to die earlier in culture and so the carriers are at risk for late onset disorders including FXTAS and FXPOI.

The FMRP is a RNA binding protein that shows preference towards RNA homopolymers as for certain subgroups of cerebral transcripts. The protein encoded by the FMR1 gene is involved in the regulation of the RNA stability, subcellular transport and translation of neural mRNAs that codify for proteins involved in the synapsis development, neural plasticity and brain development. Various studies have revealed that in the absence of this protein, a wide range of neural mRNAs are altered, augmenting neural protein synthesis and resulting in dendritic spine dysmorphogenesis and an excitation/inhibition imbalance (Glutamate/GABA), phenomena present in FXS. The dendritic spine dysmorphogenesis plays a role in the clinical manifestations of the syndrome, due to the weak synaptic connections leading to intellectual deficits and behavioral problems. Multiple neurotransmitter systems are impaired because of the lack of FMRP and there is enhanced protein production in the hippocampus and throughout the brain.

Heritability

FXS heritability does not have a Mendelian classic inheritance pattern. It depends on the number of trinucleotide CGG repeats within the promoter of the *FMR1* gene. The transition from PM to FM alleles occurs because of expansion phenomena during the transmission of the maternal X chromosome carrying PM, to her children. This expansion does not occur during the transmission of the paternal X chromosome, with the PM, to their daughters. So all daughters of men with the premutation will be obligate carriers of the premutation but then these daughters have a 50% risk to have children with FXS.

Dynamics of the Mutation

The risk of transition from PM to FM in the descendants depends on the number of trinucleotide repeats in the PM allele, reaching ~100% for PM alleles with more than 99 repeats. The expansion from PM to FM in meiosis can occur in alleles with as little as 56 CGG repeats; the odds that this occurs depends on the range of repeats in which the patient is classified, number of AGG interruptions and age of the mother. Normally there is an AGG anchor with every 9 or 10 CGG repeats in *FMR1*. The anchors can modify the risk for expansion of the CGG repeat when passed on by the mother. Women with the premutation and 2 AGG anchors have a lower risk of expansion to the full mutation compared to women with no AGG anchors.

Table: Comparison of the expansion risk from premutation to full mutation between reported percentages depending solely in the number of CGG repeats2 and recently reported percentages depending in AGG interruptions and maternal.

Range of CGG repeats in the permutated mother	Expansion risk to FM depending solely in the number of CGG repeats (%)	Age of the mother	Expansion risk to FM depending on the CGG repeats and AGG interruptions		
			0 AGG interruptions	1 AGG interruptions	2-3 AGG interruptions
55-59	3.7	20	0.4-1.5	0.1-0.5	0
		30	1.6-6.4	0.5-2.2	0-0.1
		40	6.5-23.1	2.3-9.1	0.1-0.5

60-69	5.3	20	1.5-22.7	0.5-8.9	0-0.5
		30	6.4-56.3	2.2-30.0	0.1-2.1
		40	23.1-85.0	9.1-65.3	0.5-8.6
70-79	31.1	20	22.7-84.7	8.9-64.8	0.5-8.4
		30	56.3-96.0	30.0-89.0	2.1-28.7
		40	85.0-99.1	65.3-97.2	8.6-63.8
80-89	57.8	20	84.7-99.0	64.8-97.2	8.4-63.3
		30	96.0-99.8	89.0-99.3	28.7-88.3
		40	99.1-100	97.2-99.8	63.8-97.1
90-99	80.1	20	99.0-99.9	97.2-99.8	63.3-88.2
		30	99.8-100	99.3-100	88.3-99.3
		40	100	99.8-100	97.1-99.8
>99	100	20	99.9	99.8-100	88.2-100
		30	100	100	99.3-100
		40	100	100	99.8-100

The intermediate or grey zone alleles are those that possess between 45 and 54 repeats and are proposed as precursors of PM alleles. The transition from grey zone alleles to PM alleles occur because of paternal or maternal meiosis instability.

Inheritance and Recurrence Risk

- Men: The majority of the men with FM usually do not reproduce with only 1% of them reported to have offspring. Male patients with FM and FXS have a 100% chance to pass on the premutation to their daughters so their daughters will only be carriers and usually they do not present with intellectual disabilities. There is loss of the FM in the formation of the sperm and only the premutation is passed on. All of their sons will receive the Y chromosome so they will not be affected with the *FMR1* mutation. Meanwhile, men with the PM will transmitt only the PM to their daughters and the number of repeats is relatively stable.
- Women: It's noteworthy that carriers of the PM can expand their allele to FM with odds depending on the number of repeats, number of AGG interruptions and age. The impact of the AGG interruptions is attributed to the decrease of DNA polymerase slippage in replication. Therefore, this interruption give stability in the gene's transmission but does not affect gene's transcription nor translation.
- Besides, according to the newborn's sex, the clinical characteristics are going to differ. Male patients with FM will develop mental impairment. On the contrary, daughters who inherit the FM have a 30% chance to have a normal intelligence quotient, 25% to have intellectual disability with IQ less than 70, nonetheless they can present learning deficit (60%) and emotional difficulties (70%).

Fragile X Syndrome Clinical Characteristics

The FXS cognitive and emotional phenotype will depend as well on the amount of FMRP that is produced, which depends on the number of repeats and the methylation degree of the *FMR1*.

Depending on the FMRP concentration, a clinical spectrum develops. When FMRP levels are not very low the symptoms are less severe, there's a moderate emotional and learning difficulties and a normal intelligence quotient. If the production of the FMRP protein decreases or stops, a severe cognitive deficit develops causing mental impairment.

Particular physical characteristics also depend and the production of FMRP. Thus, the 80% of the patients with FXS will have one or more of their common facial characteristics. The diagnosis is suspected in men with mental impairment, particular facial characteristics as long face, large and protruded ears, and macroorchidism; this phenotype allows the distinction of patients with FXS after puberty to the patients with mental impairment and not associated with FXS. However all individuals with intellectual disability or autism should have the fragile X DNA test because sometimes the physical features are not obvious or not present, particularly in young children.

Table: Clinical characteristics of patients with full mutation and Fragile X Syndrome.

Characteristics	Gender	
	Male	Female
Facial	Large & prominent ears (75-78%) Long face Mandibular prognathism(80% adult men) Cleft palate Macrocephaly	Large & prominent ears (75-78%) Long face Mandibular prognathism Prominent forehead
Ophthalmologic	Strabismus (8%) Refractive errors	Strabismus (8%) Refractive errors
Inner Ear	Deafness due to high recurrence of ear infection	
Neurologic	Seizures (23%) Hypotonia (children) Clonus (adults) Positive palmomental reflex	Seizures (23%) Hypotonia (children) Clonus (adults) Positive palmomental reflex
Psychiatric	Poor eye contact Attention defecit hyperactivity disorder Anxiety Repetitive motor behaviour Autistic characteristics Aggression and distress crisis	Poor eye contact
Development	Mental retardation Cognitive and language deficit(last one repetitive)	Mental retardation Cognitive and language deficit(last one repetitive)
Orthopedic	Flat feet Hyperextensibilityin the metacarpophalangeal joint Scoliosis Double jointed thumbs	

Thorax	Pectus excavatum
Genitourinary	Macroorchidism (95% of adult men)
Cardiovascular	Cardiac abnormalities (mitral valve prolapse)
Other	Obesity, cramped teeth, tall or short stature

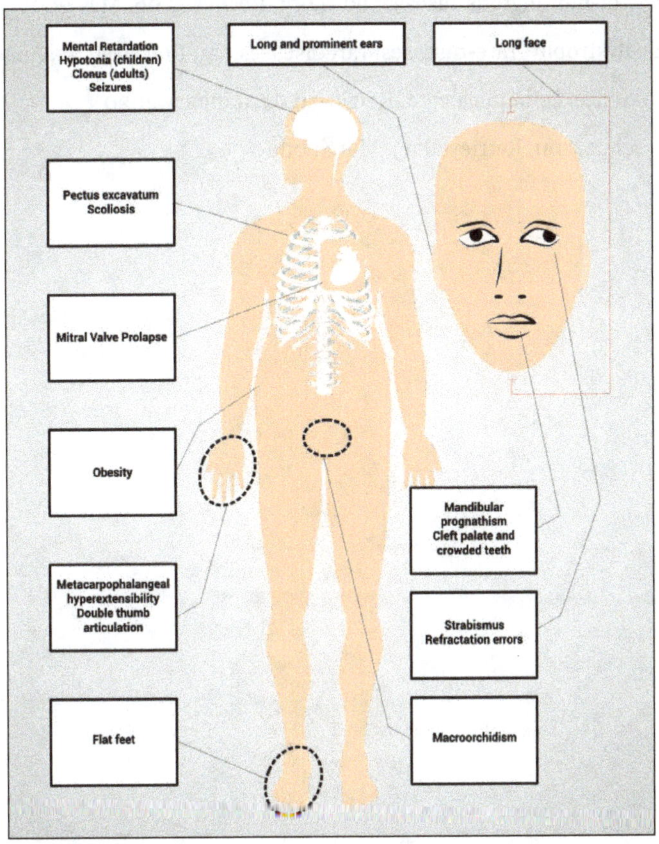

Description of the typical phenotypic characteristics of Fragile X Syndrome.

Women with FM and FXS are going to present a wider range of phenotypic characteristics than men, depending on the activation ratio of the affected X chromosome AR in peripheral blood (AR= percentage or ratio of cells with the normal allele present on active X chromosome, so that higher AR correspond to higher FMRP levels produced by the normal *FMR1* allele). The severity of the phenotype in women can vary from typical physical characteristics and mental impairment to the absence of physical phenotype together with a mild learning disability. Seventy per cent of the women with the FM present some degree of cognitive impairment.

Premutation carriers usually have a normal IQ with mild or no physical features; however, approximately 20% of women develop FXPOI or FXTAS, although 40% of male carriers develop FXTAS. Tremor/Ataxia Syndrome is a late onset progressive neurodegenerative disorder, characterized by neurological deficits that include progressive intention tremor, cerebellar ataxia, parkinsonism, neuropathy and autonomic dysfunction. Fragile X-associated primary ovarian insufficiency, which indicates the cessation of the menses before 40 years old, affects women by producing a series of complications in their fertility and reproduction, because it causes menstrual cycle irregularities, infertility and ovary hormonal deficiency.

References

- What-is-a-genetic-disorder: yourgenome.org, Retrieved 20, July 2020
- Symptoms-causes, down-syndrome, diseases-conditions-20355977: mayoclinic.org, Retrieved 12, April 2020
- Cystic-Fibrosis: physio-pedia.com, Retrieved 15, February 2020
- What-is-hd, overview-of-huntingtons-disease: hdsa.org, Retrieved 06, May 2020
- Duchenne-muscular-dystrophy, rare-diseases: rarediseases.org, Retrieved 27, August 2020
- Sickle-cell-anemia, science: britannica.com, Retrieved 04, January 2020
- What-is-thal: thalassemia.com, Retrieved 23, March 2020

Permissions

All chapters in this book are published with permission under the Creative Commons Attribution Share Alike License or equivalent. Every chapter published in this book has been scrutinized by our experts. Their significance has been extensively debated. The topics covered herein carry significant information for a comprehensive understanding. They may even be implemented as practical applications or may be referred to as a beginning point for further studies.

We would like to thank the editorial team for lending their expertise to make the book truly unique. They have played a crucial role in the development of this book. Without their invaluable contributions this book wouldn't have been possible. They have made vital efforts to compile up to date information on the varied aspects of this subject to make this book a valuable addition to the collection of many professionals and students.

This book was conceptualized with the vision of imparting up-to-date and integrated information in this field. To ensure the same, a matchless editorial board was set up. Every individual on the board went through rigorous rounds of assessment to prove their worth. After which they invested a large part of their time researching and compiling the most relevant data for our readers.

The editorial board has been involved in producing this book since its inception. They have spent rigorous hours researching and exploring the diverse topics which have resulted in the successful publishing of this book. They have passed on their knowledge of decades through this book. To expedite this challenging task, the publisher supported the team at every step. A small team of assistant editors was also appointed to further simplify the editing procedure and attain best results for the readers.

Apart from the editorial board, the designing team has also invested a significant amount of their time in understanding the subject and creating the most relevant covers. They scrutinized every image to scout for the most suitable representation of the subject and create an appropriate cover for the book.

The publishing team has been an ardent support to the editorial, designing and production team. Their endless efforts to recruit the best for this project, has resulted in the accomplishment of this book. They are a veteran in the field of academics and their pool of knowledge is as vast as their experience in printing. Their expertise and guidance has proved useful at every step. Their uncompromising quality standards have made this book an exceptional effort. Their encouragement from time to time has been an inspiration for everyone.

The publisher and the editorial board hope that this book will prove to be a valuable piece of knowledge for students, practitioners and scholars across the globe.

Index

A

Abbreviations, 41

Adenine, 4, 6-7, 10-11, 23-25, 33, 173

Adenosine Triphosphate, 6, 25-26

Allele Frequency, 75, 102, 104

Amino Acid, 4, 13, 15, 21-22, 33-34, 37-44, 51-52, 78, 100-101, 106, 173, 180, 185, 188, 202

D

Dna Double Helix, 9, 178

G

Gene Expression, 5, 9, 21, 31, 51, 105, 110-111, 116-119, 125-128, 172, 176, 178, 181-182, 188, 195, 197

Gene Product, 139, 147, 152

Gene Therapy, 120-121, 139-142, 152, 156-157, 159, 193, 195-206, 225

Genetic Code, 4, 15, 22, 33-35, 37-38, 42-43, 81, 109, 120, 130

Genetic Engineering, 120-121, 124-125, 127, 132, 137-138, 140, 195, 204-205

Genetic Modification, 135-136, 153, 155-156, 197, 205

Genetic Mutation, 48, 214

Genetically Modified Organism, 120, 138, 152-153

Genotype, 37, 44-47, 49-50, 58, 62-63, 66, 90, 95, 97-101, 107, 109-110, 133, 161, 163, 189, 192

Glutamine, 13, 23, 44

Good Manufacturing Practices, 129

Green Fluorescent Protein, 142

Guanine, 4, 6-7, 11, 23-25, 33, 38-39, 147, 173

H

Haploid Gametes, 81

Homeostasis, 79

Human Genome, 3, 121, 159, 162, 167, 171-172, 176-188, 195, 204-207

Hybridization, 9-10, 57, 71, 84, 132, 162

Hyperactivity, 231

I

Immunohistochemistry, 132

Intellectual Property, 138

Introns, 3, 5, 15, 106, 172, 177, 180-182

Ionizing Radiation, 10, 67

Isomerases, 179

L

Lactate, 130

Lagging Strand, 16

Lepidoptera, 126

Leucine, 22, 39

Loops, 12, 29-31, 41, 68

M

Meiosis, 17-18, 26, 33, 45, 48, 60-61, 64, 74, 79-81, 83, 85, 87-89, 91, 109, 140, 176, 229-230

Messenger Rna, 4, 10, 13, 105-106, 200

Metabolism, 7, 11-12, 14-15, 20, 23, 26, 79, 112, 114-115, 204

Methionine, 23

Microorganism, 153

Mitochondria, 3, 8, 15, 22, 163, 173, 176, 191

Mitochondrial Dna, 16, 162-163, 177, 188, 191

Mitosis, 80, 83, 87, 109, 131

Mitotic Spindle, 29

Molecular Cloning, 121, 123

Morphogenesis, 105

Morphology, 46, 49, 110

Myeloma, 195, 202

N

Nervous System, 78, 222

Next Generation Sequencing, 179

No-till Farming, 154

Noncoding Dna, 176, 178, 180-181

Nucleic Acid, 5-6, 12, 15, 54, 176

Nucleic Acids, 5-6, 20, 45

Nucleotide Sequence, 13, 21, 43, 54, 107, 120, 122

Nucleotide Structure, 24-25

Nutritional Quality, 134

P

Parkinson's Disease, 195

Pathogenic Bacteria, 116

Pollen Grains, 85

Polymer, 7, 12, 136, 156, 195, 200

Polymerase Chain Reaction, 132, 144

Polymers, 195
Post-transcriptional Modification, 106
Precipitation, 227
Prokaryotes, 3-4, 8-9, 13, 15, 21, 105-106, 110-112, 116-119, 196
Promoter Sequence, 130
Protein Synthesis, 4-5, 12-13, 20-22, 41, 44, 106, 109, 118, 182, 229

R
Radioactivity, 41, 71
Recombinant Dna Technology, 153, 182, 225
Redundancy, 39, 74, 148-149, 151
Resolution, 7, 170, 176
Reverse Transcription, 107, 131
Rhizobia, 83
Ribonucleic Acid, 4-6, 12, 24
Ribosomal Rna, 13, 178, 182
Ribozymes, 14
Rna Polymerase, 12, 16, 20-21, 105-106, 112-118, 181

S
Saccharomyces, 125
Sanger Method, 12
Senescence, 79
Sickle Cell Anemia, 156, 187, 195, 222-224
Signal Transduction, 107
Silent Mutations, 75
Soil Erosion, 154
Somatic Cells, 26, 83, 107, 121, 140, 142, 145, 151, 176
Source Code, 1-2
Staphylococcus Aureus, 135
String, 9, 38, 64
Substrates, 51, 114-116
Synthesis Pathway, 50

T
Time Scale, 179
Trajectories, 102
Transfer Rna, 4, 12-13, 40-41, 105-106, 178
Transferases, 179
Transporters, 179
Tryptophan, 22, 39-40, 42, 44, 112-113, 117

U
Ultraviolet Absorption, 10
Uracil, 4, 6, 11, 20, 23, 25-26, 38-40

W
Western Blotting, 132

www.ingramcontent.com/pod-product-compliance
Lightning Source LLC
LaVergne TN
LVHW080734210926
848485LV00003B/558
9781639895380